Pedestrian Malls and Skywalks

Traffic separation strategies in American downtowns

KENT A. ROBERTSON

Avebury

Aldershot · Brookfield USA · Hong Kong · Singapore · Sydney

Published by
Avebury
Ashgate Publishing Limited
Gower House
Croft Road
Aldershot
Hants GU11 3HR
England

Ashgate Publishing Company
Old Post Road
Brookfield
Vermont 05036
USA

British Library Cataloguing in Publication Data

Robertson, Kent A.
Pedestrian Malls and Skywalks:
Traffic Separation Strategies in American
Downtowns
I. Title
ISBN 1 85628 687 8

Library of Congress Cataloging-in-Publication Data

Robertson, Kent A.
Pedestrian malls and skywalks : traffic separation strategies in
American downtowns / Kent A. Robertson. -- (1st ed.)
p. cm.
Includes bibliographical references.
ISBN 1-85628-687-8 : $68.95 (est.)
1. Pedestrian areas. 2. Pedestrian areas--United States.
3. Pedestrian facilities design. 4. Pedestrian facilities design-
-United States I. Title
HE336.S5R63 1994 94-8723
388.4'1--dc20 CIP

Printed and Bound in Great Britain by
Athenaeum Press Ltd, Newcastle upon Tyne.

Contents

Figures and tables

Acknowledgements

This book represents the culmination of ten years of research and writing on the topic of downtown skywalks and pedestrian malls. While many of the chapters have appeared as articles in journals such as Town Planning Review, Journal of the American Planning Association, Transportation Quarterly, and Cities, I thought it would be very helpful to people interested in downtown pedestrian planning to have all of these articles assembled into one convenient volume.

It is impossible to thank all of the people individually who assisted me with my research. In each of the 19 case cities reported on in this book, there were numerous individuals who provided me with invaluable information, time, contacts, and perspectives. Without the help of these numerous people, the completion of this book could not have happened. I do wish to thank the Swedish Information Service (New York) and the Swedish Institute (Stockholm) for the funding they provided which allowed me to conduct my research in Sweden during 1990. I would also like to extend my warmest appreciation to Hans and Anna Gaste of Leksand, Sweden; without their hospitality, assistance, and support throughout my travels in Sweden, the Swedish research would not have been possible.

Finally, I would like to dedicate this book to my parents, Jeanne and Earl Robertson of Rochester, New York. The love, support, and encouragement they have given me throughout my life provided me with the ability, motivation, and inspiration to produce this work.

1 Pedestrians and downtowns

Pedestrians constitute the oldest and most widely utilized form of transportation in the world. While walking may not be able to match the speed, distance, and convenience provided by mechanized transportation modes such as mass transit and the automobile, pedestrian transportation is remarkably efficient and beneficial. It allows for tremendous flexibility in movement, as one is not limited to roads or transit tracks. The space requirements are minimal when compared to other means of transportation. It is relatively inexpensive. Other intrinsic rewards of walking to the individual include physical fitness, aesthetic pleasures, and opportunities for socializing (Pushkarev and Zupan 1975). Society at large benefits because pedestrian transportation results in minimal harmful impacts on the physical environment and reduced use of fossil fuels. Psychologists (Hillman 1989) have even suggested that increased walking can be therapeutic and can help to alleviate behaviorial and personality disorders.

Despite the many benefits associated with walking, its widespread use, and the fact that walking constitutes at least a part of almost all journeys (i.e., getting from one's car to one's final destination), the truth is that the pedestrian is ignored by most transportation experts. Most books on urban transportation barely even acknowledge the existence of pedestrians. Because of its ubiquity, it tends to be taken for granted. Tolley (1990) believes that walking is ignored because it does not directly lead to major problems; in other words, it does not cause pollution, accidents, noise, or congestion, so transportation experts pay little attention to it. According to William Whyte (1988), the pedestrian is frequently seen as a problem in transportation planning - getting in the way of traffic flow - and not as an an integral part of the movement system.

The plight of the contemporary pedestrian

One can safely say that the typical American today, whether in downtown, the neighborhoods, orthe suburbs, walks far less frequently than did people from previous generations. Part of the explanation is that people do not like to walk due to laziness and/or the ongoing love affair between the American public and the private automobile. However, many people choose to minimize walking because it is oftentimes difficult and unpleasant. A number of factors help to contribute to this unfriendly environment for pedestrians.

The first factor is the low priority granted pedestrian transportation in relation to the automobile. Whether in downtown or the suburbs, the needs of the automobile driver usually receive higher priority in terms of space, convenience, and safety. For example, it is often dangerous and time-consuming for pedestrians to cross busy streets, where traffic signals are almost always timed to move automobiles efficiently, not pedestrians. In addition, drivers are usually more cognizant of other cars than of pedestrians, particularly on right-hand turns on red lights. When automobiles and pedestrians have to compete for precious downtown right-of-ways, the result more times than not is that streets are widened at the expense of sidewalk; in most American downtowns, sidewalks are narrower than they were in the 1940s (of course there are no sidewalks at all on many suburban thoroughfares!). For those who are still unconvinced regarding urban transportation priorities, I pose the following question: After a heavy snowfall, are streets or sidewalks the first to be cleared?

A second factor contributing to our unfriendly pedestrian environment is the increasing distances between destinations, thereby making walking not only less attractive, but in some cases, particularly in the suburbs, nearly impossible. Even in our relatively dense downtowns, densities have decreased and distances between activities have increased because of a) the vast amount of space devoted to the automobile in the form of parking lots/ramps, b) many downtown office buildings constructed in recent years are set back from the street, include landscaped grounds, and encompass an entire city block, and c) downtown activities have expanded beyond the range of the central core as the traditional value of compactness is negated in a society based on the automobile.

The abundance of obstacles to be found on the sidewalks constitutes a third factor. The already narrower sidewalks are often littered with traffic signs, parking meters, and traffic signal poles, all of which serve the needs of the automobile. Add to this items such as telephone poles, streetlights, mailboxes, trashcans, and newspaper machines, and it becomes clear that a significant amount of sidewalk space is cluttered with objects. These objects not only take up valuable pedestrian space, but they force the pedestrian to be continually alert so as not to run into any of these obstacles, thereby reducing the overall pleasure of walking. The more time a pedestrian spends avoiding potential obstacles and dangers (i.e., traffic), the less time remains for enjoying the pedestrian experience.

A final factor concerns the overall quality of the pedestrian environment. People will choose to walk if, among other factors, the route contains high levels of interest and aesthetics in terms of people, activities, and architecture. Most commercial strips in the suburbs are void of these qualities, and therefore are virtually hostile to the pedestrian experience. Even the contemporary downtown contains far too many blocks that are uninteresting for the pedestrian to walk by. Many downtown blocks are dominated by the "blank walls" of modern office buildings (Whyte 1988), a lack of street-oriented shops, parking ramps/lots, and abandoned buildings, all of which encourage a person to walk by as quickly as possible, and not to savor the walking experience.

The downtowns of some newer cities of the sunbelt (i.e., Phoenix, Houston) are even less conducive to the needs of the pedestrian because much of their growth and development took place since the mid-20th Century (the automobile age) as opposed to the 19th and early 20th Centuries, when downtowns were compact to better serve mass transit. As a result of this later development, sunbelt downtowns are frequently characterized by a) buildings and activity centers that are even further apart, b) wider streets and narrower sidewalks, c) building fronts that are even less oriented to the street-level, and d) some streets that are not designed for pedestrian use at all.

Despite the laziness and over-reliance on the automobile attributed to Americans, people will choose to walk provided that a number of conditions are met. First, the weather must be pleasant; obviously, rain, snow, and extreme heat or cold will limit the number of people opting to walk. Second, distances between destinations must be deemed walkable; this varies depending on the individual pedestrian, setting, and trip purpose, but, for example, Houstoun (1990) estimates that downtown pedestrians will walk no further than three blocks to reach their destination. Third, the experience of walking must be perceived as enjoyable. This enjoyment can be enhanced both by the elimination of obstacles (i.e., automobile traffic, objects placed on sidewalk) and by providing a safe and comfortable environment with interesting scenery which can include people, activities, buildings, shops, landscaping, and natural vistas.

The elimination of obstacles is interconnected with the potential enjoyment of the environment: One's appreciation of their surroundings will be reduced if they have to constantly be alert for obstacles. The prototypical suburban mall provides an excellent example of this point. People will go to great lengths to park their cars as close to mall entrance as possible, often passing up empty spots located at the periphery of the parking lot. The reason for this behavior is to minimize the unpleasant and sometimes dangerous walk in the mall parking lot. However, once inside the mall with it's attractive and obstacle-free pedestrian environment, people will spend hours walking around. This lesson from suburban shopping malls can be applied to pedestrian behavior on city streets as well.

Pedestrians and the changing downtown

As the American downtown has changed considerably during the 20th Century, so too has the role of the downtown pedestrian. During the early decades of the century, the pedestrian, along with electric streetcars, dominated transportation throughout the city, and particularly in the downtown. Streetcar systems were laid out in a radial design that converged on downtown, thus bringing thousands of workers, shoppers, and others to the central core. Within close walking distance of downtown were numerous residential neighborhoods which also contributed to the high volume of downtown pedestrians. As a result, the turn-of-the-century American downtown was very pedestrian-oriented. Typically, between one-third and one-half of the right-of-ways were for walking (Pushkarov and Zupan, 1975), and it was common for the street to serve as an overflow pedestrian corridor when the walkways were crowded. Key characteristics of this downtown include high density levels, highly concentrated activity centers, streets and sidewalks pulsating with human activity, and streets lined with a continuous series of small shops.

Downtowns of today, conversely, usually are characterized by lower densities, activity centers that are more spread out, lifeless streets and sidewalks, especially in the evenings and on weekends, and a continuously decreasing amount of street-oriented shops. The sum total of these changes during the 20th Century is that our downtowns are not very inviting to pedestrians and, therefore, the volume of pedestrians on downtown streets has decreased dramatically.

A series of factors help to explain why downtowns have changed so much in terms of their pedestrian use. First, the number of transit riders in this country has declined from 16.7 billion in 1925 to 9.1 billion in 1989 (American Public Transit Association 1990). Furthermore, the majority of 1925 riders ended up downtown either as their final destination or to transfer to another line, whereas many transit riders today never set foot downtown. Transit compliments the needs of a pedestrian-oriented city because it promotes higher densities and concentrated activities. In addition, transit riders are more likely to stay downtown for multiple purposes, as opposed to the automobile driver who may choose to go elsewhere for eating, shopping, etc.

A second factor, one closely related to the first, is the large amount of downtown space devoted to the driving and storing of private automobiles.

According to Francis (1987, p. 24), 85% of city right-of-ways are devoted to moving vehicles and 15% to pedestrians. And these estimates do not include the large tracts of off-street land used for parking vehicles. The increased use of automobiles has exerted three negative impacts on downtown pedestrian use: 1) it has made crossing streets more time-consuming and dangerous, 2) it has reduced densities and spread out activities, and 3) it has littered the cityscape with ugly parking lots and "blank wall" parking ramps, which serve to decrease the enjoyment level associated with walking downtown (few people enjoy walking next to a parking lot!).

Third, the effects of the federal Urban Renewal programs from the 1950s and 1960s worked against the needs of pedestrians. Local governments, with financial backing from Washington, D.C., cleared many downtown blocks, thereby destroying thousands of small independent shops and businesses. Urban Renewal also destroyed many lower income neighborhoods located on the periphery of downtown, thus eradicating a source of human activity that helped fill downtown streets. After these buildings were cleared, the cities often could not find viable development projects to fill the spaces, leaving many downtowns with weed-infested parking lots and large gaps in the downtown fabric. One unintended but very serious impact of Urban Renewal was significantly decreases densities (Frieden and Sagalyn 1989).

One final factor involves the massive suburbanization process that has resulted in a majority of Americans who do not live, work, or shop in the downtown. One consequence of this is attitudinal, wherein suburbanites no longer feel comfortable and safe spending time downtown. The images of crime, homelessness, and drugs that they see on the evening news, coupled with their lack of familiarity with what the downtown has to offer, has created a stigma about going downtown. It is a alien environment to many suburbanites. Obviously, walking in such an environment would not be perceived by many suburbanites as a potentially desirable and enjoyable endeavor.

Despite the decreasing numbers of pedestrians to be found on downtown sidewalks, they are still indispensible to a vital downtown district. Brambilla and Longo (1977) believe that the benefits of pedestrianization to a downtown can be broken down into four categories: 1) traffic management, 2) economic revitalization, 3) environmental improvements, and 4) social benefits.

The first category concerns traffic management. A healthy core is often characterized by a balanced transportation system. By improving the quality of the pedestrian environment, and thereby increasing the number of people walking, less space will be available for automobiles downtown, which should result in less motor vehicles coming downtown and more people using mass transit (provided the city has an efficient transit system). Thus, a better balance between mass transit, pedestrians, and automobiles can be achieved.

Second, the economic revitalization of downtown can be facilitated. High levels of pedestrian traffic are required to support the remaining street-level businesses as well as the more recently opened indoor-oriented shops located off atriums and/or skywalk systems. A pedestrian-friendly downtown will encourage people to wander throughout the district, thereby walking by more establishments; conversely, downtowns without an attractive environment for walking will result in many people not venturing much beyond the office building or indoor shopping center they are presently using. And, quality public spaces that are pedestrian-friendly will intice more people to come downtown more often and to stay longer once they are there; the longer people stay, the more money they will spend downtown.

Third, increased pedestrianization will lead to an improved downtown environment. More pedestrians and less automobiles would result in significant reductions in air and noise pollution, both of which are usually high in downtowns. For instance, noise levels were cut in half on Stroget, Copenhagen's (Denmark) famous pedestrian street, when automobiles were eliminated (Brambilla and Longo, 1977, p. 22). Pedestrianization can also enhance the historic environment of downtown, a justification used frequently in Europe. Conservation is served by reducing the levels of automobile emissions that damage older buildings and by encouraging the appreciation of the historical structures by allowing them to be viewed without the interruption of automobile tops and traffic signals/signs.

Fourth, numerous social benefits are possible when the downtown environment is improved for pedestrians. Opportunities for socialization and interaction with a wide range of people are certainly greater for pedestrians than for automobile drivers. Since downtown still attracts a heterogenious mix of people, the chances of being exposed to persons with different cultural, economic, and racial background are better than in most other settings, a characteristic especially important for children. Pedestrian-oriented open spaces with quality seating are particularly welcomed by certain segments of the population such as the elderly, children, the handicapped, and parents with infants. A person will probably feel safer walking on a downtown street full of pedestrians than they would on a street where they were all alone.

Increased pedestrianization is associated with one additional benefit - the vitality and image of a downtown is often gauged by the level of pedestrian activity on the street. Even though other indicators may more accurately reflect the economic health of downtown, the average person determines for themselves whether a downtown is lively or dead by the volume of pedestrians, the level of which is largely determined by the quality of the pedestrian environment. As stated by Zimmerman (1974, p. 47), "....ultimately, it is as pedestrians that (people) capture the urban experience, and the greater the diversity and ease of contact the city offers pedestrians, the more likely it is to attract and hold them."

Pedestrian planning strategies

It is clear that the role of the pedestrian has declined in most American downtowns and yet pedestrianization continues to be important to the survival of downtown. Given this paradox, what can city planners and urban designers do to help improve the downtown pedestrian environment? Some of the more commonly used strategies are presented below:

Widen sidewalks

As stated above, downtown sidewalk widths have been reduced over the years in order to increase the street capacity for automobiles. For example, New York City's minimum standard sidewalk width decreased from 15 feet in 1912 to 11 feet in 1963 (Brambilla and Longo, 1977, p. 25). A staightforward solution is to widen sidewalks, thereby allowing the pedestrian to reclaim lost space (Breines and Dean, 1974). However, advocates of the automobile, including a majority of downtown merchants, would strongly oppose this strategy because of the fear that it would lead to increased traffic congestion and thus prevent some drivers from venturing downtown.

Discourage automobile traffic

The idea behind this somewhat radical strategy is that if public policies could be implemented to make it more difficult and inconvenient to drive downtown, then the downtown could gradually become more pedestrian-oriented. Parking policies are one possible method. By reducing the total number of parking spaces, raising the cost of parking, and/or placing more restrictive time limits on parking, a city could effectively encourage drivers to seek alternative means of transportation to downtown (i.e., carpool, transit). Of course, for this to be successful, complimentary policies are needed to promote/coordinate ridesharing and improve the quality and capacity of public transit.

Climatization of the pedestrian environment

Inclement weather can inhibit people from choosing to walk. Climatization of the downtown shopping experience in the form of indoor gallerias has already proven to be immensely popular. Extending this concept to the pedestrian environment, a number of cities have constructed climate-controlled pedestrian/shopping corridors either below or above (skywalks) the street-level. Other climatization techniques include heated sidewalks for melting the snow (i.e., Nicollet Mall in Minneapolis) and canopies to provide welcome protection from the sun or rain.

Improve safety and security

An environment which is perceived to be dangerous serves as a major impediment to pedestrians. This is especially true for downtowns, an unfamiliar and risky environment in the minds of many citizens, particularly suburbanites. Strategies used to make areas appear to be more safe and secure include improved lighting and increased foot patrols by police officers. Ironically, the presence of automobile traffic actually increases one's sense of security, especially where there are few pedestrians; therefore, on pedestrian-only corridors, it has been recommended there be a cross-street with vehicular traffic at least every 100 yards (Antoniou 1982).

Increase attractiveness of walkways

The principle at work here is quite simple: Attractive and activity-laden walkways will promote increased pedestrian volume. Benepe (1965, p. 34) refers to this as the "delight" factor; features that can improve the "delight" of walking include colorful storefronts, outdoor exhibits, attractive landscaping, and chance meetings, and anything else that results in the discovery of something new or vital in one's surroundings. People-watching is also a desirable activity for many pedestrians. Therefore, high pedestrian volumes will attract even more people to that particular walkway. To increase activity levels along pedestrian corridors, planners have encouraged kiosks, fruit and vegetable stands, sidewalk cafes, outdoor eating, and artists and street performers.

Increase quantity and quality of seating spaces

Quality pedestrian spaces are characterized not only by interesting and active environments for walking, as discussed in strategy #5 above, by also by availability of seating throughout the pedestrian corridor. If adequate seating is provided for people to sit, rest, talk, and people-watch, then the total volume of people using the pedestrian space will increase considerably. In other words, people will stay longer if quality seating is available. Seating is welcomed by everyone, but is particularly needed by the elderly, children, and parents with infants. Some important design considerations for quality seating are orientation towards the sun, quality of view (usually facing the heaviest pedestrian flows is best), and weather protection (Brambilla and Longo 1977). Besides the traditional bench, other forms of seating include movable chairs, wide ledges, and steps.

Change existing zoning ordinances

An often unintentional consequence of many current

zoning ordinances is the creation of an unfriendly pedestrian environment. Several types of changes can help this situation. First, as suggested by numerous observers (i.e., Breines and Dean 1974, Brambilla and Longo 1982, Untermann 1990), single-use zoning is inconsistent with the requirements for a healthy pedestrian district. Mixed-use zoning (i.e., encouraging more residential activity in downtowns dominated by commercial uses) will lead to a fuller spectrum of urban uses and users, and will extend the time periods during which the pedestrian district is active. Second, zoning that encourages deep setbacks from the sidewalk serves to interrupt the continuous streetwall of urban activities which are so vital to providing an interesting walking experience. Third, zoning that requires high parking ratios works to serve the needs of the automobile and usually results in an unfriendly pedestrian environment. As discussed in strategy #2, limiting the number of downtown parking spaces may encourage more people to walk.

Separate vehicles from pedestrians

A primary goal of downtown planning for many years has been to separate vehicular from pedestrian traffic, supposedly for the benefit of both. The basic assumption has been that both modes of transportation will flow more smoothly and safely if the instances of intersection can be minimized. Some observers, most notably Whyte (1988), have argued that vehicular traffic benefits to a far greater degree because the automobile receives the prime street-level space, while the pedestrian is often relegated to above-grade or underground corridors. Nevertheless, downtowns that have separated the two forms of transportation usually can claim to possess a pedestrian-friendly environment in which the pedestrian does not have to worry about continually being alert for automobile traffic. The three majors methods of separation are underground concourses, most of which are linked with shopping complexes and/or underground transit (i.e., Houston, Montreal, Rochester MN), above-grade skywalk systems, and grade-level pedestrian malls. European cities have used underground concourses and pedestrian malls most frequently to help pedestrianize their city centres; the use of overhead skywalks is virtually unknown in Europe. Skywalks and pedestrian malls are much more common in American downtowns than are underground concourses, and will constitute the focus of the remainder of this book.

Purpose and scope of book

The purpose of this book is to examine in detail the two most popular forms of downtown traffic separation mentioned above - skywalks and pedestrian malls. The main foci will be on the characteristics and problems of skywalks and malls, with particular emphasis being placed on design, transportation considerations, commercial activity, and usage patterns. Furthermore, the extent to which these planning strategies both facilitate and hinder the objectives of downtown redevelopment will be assessed. The major focus of the book will be on how these pedestrian separation strategies have worked in American downtowns. However, an analysis of Swedish pedestrian malls and a detailed case study of a Canadian skywalk system are also included for the purpose of comparing American downtown pedestrian systems with those in other parts of the world.

My conclusions are based on a series of case studies conducted between 1985-1990. In 1985 and 1986, downtown skywalk systems in Calgary (Canada), Cincinnati, Des Moines, Duluth, Minneapolis, and St. Paul were evaluated. Pedestrian malls in Denver, Madison (Wisconsin), Minneapolis, Providence, St. Cloud (Minnesota), and Santa Cruz (California) were studied during 1988. For a comparative perspective, pedestrian streets in six Swedish cities were examined in 1990: Falun, Gothenburg (Goteborg), Linkoping, Orebro, Stockholm, and Uppsala. Each city was visited by the author for 3-5 days for the purpose of 1) making observations of the pedestrian system's features, design, use, and problems, 2) taking inventories of economics activities along the mall or skywalk, and 3) conducting key informant interviews with well-informed planners, city officials, and businesspeople. For the skywalk cities, a survey of users was also administered. The methodology used and the issues examined were similar for both mall and skywalk cities, thereby enabling valid comparisons to be drawn regarding economic impacts, design issues, image, transportation considerations, and the relationship to overall downtown development goals.

Listed below are a series of key questions which the following chapters will address:

1. What types of cities have implemented skywalks and pedestrian malls (population size, geographic region)?

2. From city to city, what are the differences between pedestrian malls (or skywalks), and why do these differences exist?

3. Why are many pedestrian malls in trouble in the United States, while they appear to be thriving in European cities?

4. What has been the impact of indoor suburban shopping centers on downtown skywalks and pedestrian malls?

5. How do the two pedestrian-separation strategies relate to the automobile and mass transit in downtowns?

6. To what extent do skywalks and pedestrian malls change downtown, both positively and negatively?

7. What is the impact of these two strategies on downtown commercial activity?

8. How important is urban design to the success of these strategies?

9. What is the impact of these pedestrian-separation strategies on downtown activity patterns?

10. What is the future for skywalks and pedestrian malls in downtowns?

11. What factors should a city take into consideration when deciding which strategy would work best in their downtown?

Review of related literature

Separate bodies of literature have developed for pedestrian malls and skywalks. Research and discussion on pedestrian malls was abundant in the

1960s and 1970s, with much of the work focusing on design features and attempting to draw comparisons between Europe and the United States (for example, Wolfe 1962, Contini 1969, Brambilla and Longo 1977, and Rubenstein 1978). Malls were also discussed and evaluated in more general books on pedestrians (Rudofsky 1969, Breines and Dean 1974, Pushkarev and Zupan 1975) and downtown planning (Gruen 1964). Interestingly, both Rudofsky and Gruen were very critical of many American pedestrian malls because of their tendency to be implemented and viewed as a single design element, not as a contributing part to the larger downtown environment.

As mall construction declined in America, so did the volume of corresponding literature. In recent years, however, several comprehensive studies have been published which examine the evolution of pedestrian mall development in one country, such as Denmark (Berdichevsky 1984), West Germany (Hadju 1988, Monheim 1990), the United States (Robertson 1990), and Sweden (Robertson 1991). Comparative works include Hass-Klau (1990), who does an excellent job of analyzing the development of pedestrian streets and other pedestrianization schemes in Britain and Germany, and Robertson (1992), who compares downtown pedestrian streets in Sweden and the United States.

The literature on skywalks has focused exclusively on North America because they are not yet a common feature of European townscapes. Much of the early literature was descriptive and non-critical (for example, Fraser 1970, Fruin 1973, Podolske and Heglund 1976). The 1980s has produced a body of work which is more analytical and critical, such as Forusz (1981), Jacob and Morphew (1984), Dillon (1985), Robertson (1988), and Maitland (1992). The popularity and problems of skywalks as a development tool has been reported in both the popular press (Anderson 1988, Belkin 1988) and in professional magazines (Morphew 1984, Lasser 1988). Whyte (1988), in his widely read book on urban public spaces, devotes a chapter to skywalks in which he argues that they assist in the "dullification" of downtown.

Pedestrianization in general has been the subject of a number of recent noteworthy books, the focus of which have been the European experience. This suggests that there exists considerably more interest in pedestrian planning in Europe than in North America. Gehl (1987) discusses the pedestrian in the context of designing well-used public spaces, particularly in Scandinavia. Hass-Klau (1990), mentioned above, analyzes and compares approaches to separating vehicular from pedestrian traffic in Great Britain and West Germany. An anthology edited by Tolley (1990) presents a series of interesting articles on the prospects and problems of pedestrian and bicycle transportation in Europe; of interest, Untermann's (1990) chapter in this European-oriented book discusses why planning for the pedestrian is difficult in the United States. It is important to note that none of the literature reviewed attempts to compare the two most commonly used methods of traffic separation in American downtowns: skywalks and pedestrian malls.

2 The evolution of pedestrian malls in Europe and the United States

The impetus for constructing pedestrian malls in European city centres began during the rebuilding efforts following the destruction caused by World War II, although a few pedestrianization schemes had been attempted prior to that time. In America, there were virtually no downtown pedestrian-oriented strategies implemented until that late 1950s, when the first pedestrian mall was built in Kalamazoo, Michigan. With the help of secondary sources, most notably Brambilla and Longo (1977) and Hass-Klau (1990), the purpose of this chapter is to trace the evolution of pedestrian malls in both European and American downtowns.

Early pedestrianization schemes in Europe

The idea for separating pedestrian and vehicular traffic can be traced back to the Nineteenth Century. In London, arcades were proposed which would provide mid-block glass-covered walkways connecting busy destinations, such as railroad stations and commercial districts. One of the better known proposals was the Crystal Way (1870s) that connected Fleet Street, the Strand, and Trafalgar Square, and served as a model for the Crystal Palace in Ebenezar Howard's Garden City plan at the end of the century. In both England and France, proposals were put forth for building special pedestrian thoroughfares on terraces above street-level. Le Corbusier proposed an even more radical traffic separation plan in the 1920s that included four levels of streets, each designated for a different mode of transportation. The problem with all of these utopian proposals was that they were too costly to implement practically, and therefore they had little immediate impact on the realities of downtown traffic (Hass-Klau 1990).

During the 20th Century, many European cities began to close individual narrow medieval streets to wheeled traffic because these streets did not have the capacity to handle the needs of both pedestrians and motor vehicles. This was such a common occurrence in city centres across the continent that there exists little documentation mentioning it (Hass-Klau 1990, p. 23). The combination of the rapidly increasing use of private automobiles and medieval street systems that could not handle motorized traffic very effectively led to the widespread application of street closings and street restrictions in European city centres. These closed streets, although not intended to be pedestrian malls per se, clearly constitute a key step in the overall evolution of the mall.

Pedestrian malls in Britain, Germany, and Denmark

It is difficult to generalize about the evolution of European pedestrian malls because the timing and form of mall development differed for individual nations as a result of their unique historical, cultural, and political environment. Great Britain, West Germany, and Denmark have been selected for more detailed discussion because they represent nations with significantly different cultural values, political/economic structures, and 20th Century historical influences. For each country, an individual well-known pedestrian mall will be presented in detail. Furthermore, an in-depth case study of Swedish pedestrians streets is the focus of chapter 4.

Germany

West Germany was one of the first European nations to make widespread use of pedestrian streets in their city centres. The major reasons were 1) the major rebuilding efforts required for war-damaged city centres, particularly in northern German cities, which provided planners with an opportunity to rethink the role and function of the core, and 2) the large road-building projects which often rerouted traffic patterns so as to render narrow core streets as less important and useful to the needs of the automobile. It was clear that the creation of pedestrian streets was merely a by-product of this extensive highway building coupled with narrow medieval streets in the core which were incompatible with motor vehicle use (Hajdu 1988, Monheim 1990). Simply stated, "..medieval street patterns have tended to protect the pedestrian from wheeled traffic simply because many streets were too narrow for the passage of wheeled vehicles" (Hass-Klau 1990, p. 251).

The first generation of German pedestrian malls, according to Hadju (1988), were constructed during the late 1940s and 1950s. Kiel, Bonn, Essen, Colongne, and Wilhelmshaven were among the earliest cities to close core streets to motor vehicles. Management of city traffic flow and preservation of downtown retailing constituted the two primary factors underscoring pedestrian streets during this phase. The typical pattern for most first generation pedestrian malls was a central pedestrian-only street or district that focused on the historic core of the city - oftentimes the main cathedral - and a ring road which was placed on the former site of the medieval wall. Most of the first generation cities are located in northern Germany because war damage was more extensive and post-war industrialization progressed at a greater pace than was the case in Bavaria and other southern regions. By 1960, 35 German cities had pedestrianized central streets (Hajdu 1988, p. 325).

West Germany's second generation of pedestrian streets were built between the early 1960s and mid-1980s. By this time period, the objectives had evolved "...to enhance the environmental appeal of city centres, to preserve historical monuments and buildings by integrating them into traffic-free zones, and to encourage the city centre's role as a social meeting place" (Hajdu 1988, p. 330). More emphasis was placed on design amenities, landscaping, and activities that would attract people. Driven by the economic success experienced by first generation malls and by the new emphasis on improving downtown's physical, historical, and social environment, the second generation witnessed a tremendous proliferation of pedestrian malls. By

1971, approximately 370 German cities possessed downtown pedestrian malls (Hass-Klau 1990, p. 204), a tenfold increase from ten years earlier.

One of the most popular and well-known downtown pedestrian districts is in Colonge on the Rhine River, the fourth largest city in what was West Germany. Colonge was devastated by the war, as over 80% of its building were destroyed; remarkably, the world-famous Cologne Cathedral suffered only minimal damage from the bombings, and was able to serve as the centerpiece for post-war rebuilding efforts. When they rebuilt the city centre following the war, they retained the narrow medieval street pattern and restored many of the buildings. While the reconstruction was occurring in the 1950s, several pedestrianization schemes were introduced, such as closing smaller streets to traffic and restricting traffic on major shopping streets at certain times of the day. The early 1960s witnessed Schildergasse and Hohe Strasse - the two major shopping streets - being converted into permanent pedestrian malls. These streets connected to a nicely-designed public square in front of the great gothic cathedral and to the railroad station; the resultant pedestrian district is almost always active with shoppers and tourists, and is considered to be "among the most successful traffic-free streets in Europe" (Brambilla and Longo 1977, p. 73).

Despite the widespread application and apparent popularity of Germany's pedestrian shopping streets, they were subject to numerous criticisms. First, the success led to increased rents for commercial space, thereby forcing many small independent shops to leave in favor of larger retail chains who could afford the higher rates. This altered the composition, diversity, and overall "feel" of the pedestrian district. Second, as a consequence of the first problem, critics accused pedestrian malls as being monotonous, unimaginative, and looking pretty much like any other pedestrian mall (Hass-Klau 1990). Third, while people flocked to the pedestrianized districts, other non-pedestrianized areas of the city centre (i.e., parallel streets to the traffic-free corridor) experienced significant economic losses (Monheim 1990). Finally, the closing of selected streets to traffic often caused overloading of motor vehicles on parallel and adjacent streets.

Today, it is estimated that over 1000 German cities and towns contain pedestrian streets in their core, including virtually all municipalities with over 50,000 population (Monheim 1990, p. 247). Pedestrian malls are considered to be such a well-established and proven strategy in city centre planning and development that "a town without representative pedestrian areas now appears hopelessly antiquated" (Monheim 1990, p. 245).

Great Britain

Although the notion of pedestrianizing shopping streets in Great Britain can be traced back as far as 1923, the actual implementation of this strategy has been far less prevalent when compared to other Western European countries. There were very few pedestrian-oriented schemes for British city centres during the 1940s and 1950s. British planners were aware and apparently interested in pedestrian mall developments in West Germany and the Netherlands during this time period, but most local officials demonstrated little interest in these progressive ideas. Furthermore, British law did not permit the closing of motorized streets for pedestrian use, thereby significantly reducing the planning options available to local officials. One notable exception was Coventry, the downtown of which was totally destroyed during World War II. In 1953, Coventry rebuilt its city centre according to a new plan that included several traffic-free precincts.

In 1967, the Road Traffic Regulation Act was passed which made it legal for cities to eliminate motor vehicles from select roadways (Hass-Klau 1990). Shortly after the passage of this act, several cities created pedestrian-only streets in their downtown, most notably Norwich (1967), Reading (1968), and Leeds (1970). During the 1970s, the number of British cities implementing pedestrianization schemes increased significantly. Some of the schemes allowed buses on pedestrian streets (i.e., Manchester, Oxford, Reading).

The first British city to eliminate automobiles from existing downtown streets was Norwich, the largest city in East Anglica. Interestingly, a ruptured sewer line caused the city to close London Street, the city's principle shopping street, temporarily to motor vehicles in 1965. Pedestrian traffic and retail sales increased dramatically, surprising downtown shopkeepers and city officials. Norwich made the street closing permanent in 1967 when the Road Traffic Regulation Act was passed.

In the next few years, most London Street merchants saw retail sales increase between 5-20 percent. This strategy enabled the city to work towards its two primary planning objectives: 1) to preserve its architectural heritage, and 2) to establish the city centre as the commercial core of East Anglica. In later years, Norwich expanded its pedestrianization scheme to include several other streets in the city centre. (Brambilla and Longo 1977, pp. 99-102).

It has been estimated by Roberts (1990) that there are approximately 2000 pedestrianized streets in Great Britain today, although only a few cities (i.e., Glasgow, Norwich, Nottingham) feature pedestrian malls as central components of their downtowns. When compared to Western Europe in general, and to Germany, the Netherlands, and Denmark specifically, Great Britain has not witnessed the widespread use and popularity of pedestrianization schemes (Pharaoh and Russell 1991). Several factors help to explain this. First, British cities were less damaged by the war than were their German counterparts, and thus rebuilding downtowns was less of a priority. Second, British legislation enabling road closings was not passed until 1967, whereas most other nations had this ability since the 1940s. Third, British city centres often lacked the medieval cores so instrumental in pedestrianization efforts on the continent. And finally, widely-used urban transit systems, a key dimension to the popularity of pedestrian schemes elsewhere, were less developed in Great Britain (Hass-Klau 1990).

Denmark

The application and popularity of pedestrian streets in the socialist nation of Denmark approaches that of West Germany, despite its vastly different economic and political system. In fact, Berdichevsky (1984, p. 15) believes that "in no other country outside of Denmark has the centrally located pedestrian shopping street achieved such widespread public acceptance and recognition as a standard element of town planning." Two factors which help to achieve this high level of public support are 1) city centres that still possess their original street patterns and a large supply of sound historical structures and 2) a deeply ingrained Danish preference for human scaled townscapes

(Berdichevsky 1984).

Many Danish towns and cities were concerned in the 1950s and 1960s when more and more Danes began to drive automobiles, thereby causing congestion and damaging the physical and social environment of city centres. Aalborg, located in northern Denmark, proposed a pedestrianization plan in 1956 which was in part based on Victor Gruen's plans for the downtown of Ft. Worth, Texas. Aalborg and Copenhagen became the first Danish cities to create downtown pedestrian streets in 1962. By 1977, thirty nine cities in Denmark possessed pedestrianized commercial shopping streets (or "Gagades" in Danish), most of which utilized the original street layouts that existed from past centuries (Berdichevsky 1984).

Undoubtedly the most famous pedestrian street in Denmark is Copenhagen's Stroget, which was closed to motor vehicles in 1962. In Danish "stroget" means to stroll, a popular custom in Denmark. The Stroget actually consists of three contiguous streets which are linked to several popular squares, and contains the major shopping district in Copenhagen (i.e., major department stores, boutiques, souvenir shops, restaurants). After pedestrianization, sales increased on Stroget by 30 percent (Berdichevsky 1984, p. 17) and levels of air and noise pollution declined noticeably. Studies conducted by Gehl (1989) demonstrated that the pedestrian street was filled to near capacity in 1968 and that people did more than just walk on Stroget, they were standing, sitting, and lingering, thus raising the overall levels of human activity considerably. Eventually, Copenhagen's core pedestrian network was expanded beyond Stroget, and now includes more than 2 miles of traffic-free streets.

Stroget has served as the model for pedestrianized shopping streets throughout Denmark. Today, the goals of the majority of these pedestrian streets are to 1) reduce automobile congestion in city centres, 2) stabilize downtown commercial activity, 3) make city centres more pedestrian-friendly, especially for women, children, and the elderly, and 4) facilitate preservation of historical buildings and districts. According to Gehl (1989), who has written extensively on open spaces in Scandinavia, the fact the Danish pedestrian streets have been implemented gradually, inexpensively, and have been supported with public, not private, funding has contributed to their overall success and public support. The only major criticism of Danish gagades, according to Berdichevsky (1984), is that many have become "consumer corridors" specializing in selling luxury items to the wealthy, thereby forcing small independent shopkeepers to move elsewhere.

Early pedestrian malls in the United States

Pedestrian mall development occurred later in the United States when compared to European countries such as the Netherlands and West Germany. During the 1960s, however, retail merchants in American downtowns were being threatened by the common enemy: the suburban shopping center. These shopping centers offered ample free parking, an increasingly wide selection of goods and services, a shopping environment perceived to be pleasant and safe, and a convenient location for a rapidly suburbanizing population. In an attempt to maintain downtowns as unchallenged leaders of metropolitan retailing, city officials desperately began to seek strategies to prevent downtown businesses and consumers from fleeing to suburbia. One of the most widely implemented planning strategies was the pedestrian mall. Therefore, downtown revitalization, particularly focusing on the retail sector, was the underlying objective for the construction of most early pedestrian malls in the United States.

Coupled with this need to revive declining downtowns was a powerful ideological movement within city planning that advocated a return to human-scaled and active streets as the key to achieve successful urban renewal in both commercial and residential districts. Influential writers such as Jane Jacobs, Lewis Mumford, Bernard Rudolfsky, Victor Gruen, and William Whyte articulated the negative impacts of automobiles, highways, large-scale urban renewal projects, and single-use districts on the social and economic fabric of American cities. They argued for streets, districts, and neighborhoods that were designed to facilitate human interaction. In downtown retail districts, the implementation of pedestrian malls was consistent with this viewpoint that was gaining widespread support among many professional planners.

In the context of American cities, a pedestrian mall can be defined as a downtown corridor - usually a few linear blocks along the traditional main shopping street - where pedestrian transportation is given highest priority. Three major types of malls were built in American central business districts:

1. **Traditional Pedestrian Streets** are designed exclusively for pedestrian use, where no automobiles or transit vehicles are permitted. Even on cross streets where vehicular traffic is allowed to cross the mall, the pedestrian usually has the right-of-way. On traditional pedestrian streets, the pedestrian is literally the "king of the road". A vast majority of the malls constructed during the 1960s and 1970s were exclusively for pedestrians. Of the 94 malls presented in Rubenstein's (1978) compendium, 77 (82%) were traditional pedestrian streets.
2. **Shared Malls** (also referred to as "semi-malls" in the literature) are predominantly pedestrian thoroughfares on which limited automobile traffic at reduced speeds is allowed. The space allocated for vehicles usually is limited to one lane of one-way traffic, perhaps with an extra lane for parking. On a shared mall, access to and visibility of mall establishments are improved, but often at the expense of pedestrians who are deprived of space, aesthetics, and environmental quality. Shared malls are consistent with the concept of traffic calming used on many commercial and residential streets in northern Europe, where pedestrians share the road space with motor vehicles (see Pharaoh and Russell 1991). In recent years, a number of American cities have converted their traditional pedestrian streets into shared malls.
3. **Transit Malls** are pedestrian-oriented streets on which transit vehicles, usually buses, also are permitted. Private automobiles are not allowed, although in a few cities taxis can use the mall. Transit vehicles help to increase activity levels and to facilitate transit-pedestrian linkages, but they also can lead to negative impacts (i.e., pollution, noise) with regard to the pedestrian environment. Cities that employ transit malls include Chicago,

Philadelphia, Vancouver, Portland (Oregon), Minneapolis, and Denver.

Three distinct phases of American mall development have been identified (Brambilla and Longo 1977, Brooks 1986). During the first phase (1959-1965), the single purpose of pedestrian malls was to revitalize downtown retail activity. The planning and design of malls was heavily influenced by the downtown business community, which was primarily concerned with increased retail sales, not improving the overall quality of downtown life. Design amenities were limited to features that enhanced beautification, not pedestrian comfort. For example, while landscaping and fountains were common on early malls, many had limited seating, thus saying to pedestrians "shop or leave, but do not stay". The ability for malls to encourage social interaction was not seen as important during this first phase.

The second phase of downtown mall development took place between 1966-1970, and better incorporated the ideas of Jane Jacobs, Victor Gruen, and others mentioned above. The idea that streetscape design could improve social interaction and levels of consumer spending was embraced by planners during this phase. Thus, a wide array of activities for pedestrians was provided on downtown malls, including playgrounds, increased seating areas, and landscaping that encouraged people to stay and use, not merely look. The new approach to mall development was based on the idea that a well-designed, human-scaled downtown environment would encourage people to linger, stay downtown longer, and, ultimately, spend more money. The notion of creating malls to be linear urban parks, particularly popular in cities with milder climates (i.e., California), gained support during the late 1960s.

The final phase of pedestrian mall development encompasses the 1970s and 1980s. In addition to stimulating retail trade and improving the quality of the downtown environment, the new objective of pedestrian malls was to enhance the image of downtown and to bolster civic pride. Planners attempted to encourage greater community participation in the downtown by building open spaces on malls to accommodate concerts, festivals, speakers, and other civic events. For example, Eugene incorporated an amphitheater into its pedestrian mall. Many malls implemented during this phase sought to recapture the civic significance of the traditional "main street" by creating a pedestrian mall that would become a social, cultural, and economic centerpiece for the community. An analysis of the malls listed by Brambilla and Longo (1977) and Rubenstein (1978) reveals that more malls built in the 1970s than during the first two phases. Of the 70 pedestrian malls listed in Brambilla and Longo, 42 (60%) were constructed between 1971-1976, while 65 of the Rubenstein's 94 malls (69%) were implemented between 1971-1977 (see Tables 2.1 and 2.2).

Two early pedestrian malls that merit further discussion are located in Kalamazoo (Michigan) and Fresno (California). The first American downtown pedestrian mall was built in Kalamazoo in 1959, which had a population of approximately 85,000 at the time. Three years earlier, when the downtown was still relatively strong, concern for increasing suburban retail competition led a civic group to hire Victor Gruen's architectural firm to propose a plan to help keep the downtown prosperous. Gruen's master plan proposed a reconfigurated central transportation system that called for large-scale traffic separation, including a traffic-free downtown pedestrian precinct. City officials discarded most of Gruen's plans but decided to keep one component of his pedestrian precinct concept by closing three blocks along Burdick Street. The city was so proud of their new pedestrian mall that the city nickname was changed officially from "Celery City" to "Mall City", a name that is still used today. The mall was extended one block in 1972 and included a newly-built multi-use center. Both the original mall and the later extension were funded by assessments charged to downtown building owners and merchants (Knack 1982).

Fresno's pedestrian mall was also heavily influenced by Victor Gruen. He proposed a even more elaborate master plan for Fresno than he did for Kalamazoo. His system of transportation-separation in the core included a series of highways rings surrounding downtown Fresno which were to serve as "lines of defense" against invading automobiles. Each inward ring was more restrictive to automobile, with the core of Fresno being reserved exclusively for pedestrian use (Gruen 1964). A reconstructed Fulton Street, the major retail corridor, was to become Fulton Way, the heart of Gruen's pedestrian network. Although Fresno city officials did not adopt all of Gruen's transportation-separation ideas, they did choose to pedestrianize Fulton Street and several other cross-streets during 1964-1965. Fresno's Fulton Way possesses many of the attributes of second generation pedestrian malls because its purpose was to draw people downtown through user-friendly design (i.e., comfortable and well-placed seating, shade trees, attractive paving). Fulton Way even included two playground areas, a feature more commonly associated with malls built during the 1970s. It was one of the first malls to place emphasis on both commercial prosperity and human comfort and interaction (Brambilla and Longo 1977, pp. 141-144).

Ironically, the model for pedestrian mall design was what eventually proved to be its prime competitor: the suburban shopping center. The success of Victor Gruen's Southdale (1956) in suburban Minneapolis and other suburban malls was closely observed by downtown merchants and planners. When the first downtown malls were built a few years later to compete with suburbs for the retail dollar, they borrowed the concepts of pedestrian-only streets (the indoor corridors) and attractive design amenities (i.e., landscaping, fountains) from the suburban mall. However, the suburban mall also had climate control, healthy anchor stores, ample free parking, and a polar orientation (strategically-located anchors to ensure traffic flow throughout the mall), features that most downtowns could not replicate (Cunningham 1977).

Since 1959, when the first mall opened in Kalamazoo, more than 200 American cities have implemented pedestrian malls as a downtown redevelopment tool. Malls appear in every region of the country, with the East and Midwest being the most represented regions (see Tables 2.1 and 2.2). It is interesting to observe that this planning strategy has been more commonly applied in downtowns of smaller and medium-sized cities than in America's largest cities. An analysis of Rubenstein's (1978) list of 94 malls reveals only a handful of the country's most populated cities containing a pedestrian mall: Baltimore, Dallas, and Philadelphia (not included in his compendium are Chicago's State Street Mall and Denver's 16th Street Mall). There are no pedestrian malls in New York City, although a proposal to ban automobiles for stretches of Madison Avenue were seriously contemplated during the 1970s.

Brambilla and Longo (1977) believe this trend is unfortunate because malls are most needed in large city downtowns where traffic congestion and retailing decline is most severe. The fact that the scale of pedestrian malls is more suited to smaller cities and that smaller cities have less complicated and less congested downtown traffic patterns may partially explain why only a few large cities have adopted this strategy.

The degree to which this planning strategy can be labeled a success is still open to debate. Although urban design is very important to the ultimate success or failure of a mall, other factors must also be in place for them to work (Brambilla and Longo 1977). First, the timing of the mall is important. For example, if the mall is built after much of the retailing has already deserted downtown, it is probably doomed for failure. Second, a sound organizational structure - either private and/or public - must be in place to actively coordinate, promote, and manage the mall after construction. For example, the Denver Partnership, discussed in chapter 3, has played a key role in the success experienced by the 16th Street Mall through organizing events and promoting the mall and its activities. Finally, malls only work when coordinated with other downtown development strategies (i.e., transportation, tourism, housing, open space) to improve the desirabilty of venturing downtown and to increase overall downtown activity levels. As Cunningham (1977, p. 159) stated: "Malls are not panaceas. At best they are catalysts to future development."

Two other factors appear to be crucial when determining the success of pedestrian malls. One is the overall economic health of the downtown. Based on her study of 10 pedestrian malls, Wirtenberg (1981) concluded that cities with economically hobbled central business districts were unlikely to possess successful downtown malls. Again, for a mall to work, it must be placed in a context that includes a variety of stable economic functions and traffic generators to provide a steady supply of potential pedestrians. A second helpful factor is to have a nearby source of pedestrian traffic such as a college campus, medical center, or government center. For instance, many of malls perceived to be faring very well in the United States are in college towns such as Charlottesville (Virginia), Boulder (Colorado), and Burlington (Vermont). Madison's

(Wisconsin) renowned State Street Mall links the state government complex the University of Wisconsin campus.

While the evidence is not clear-cut, the recent perception amongst urban planners and city officials is that most pedestrian malls have failed, as witnessed by the Wall Street Journal headline stating "City Pedestrian Malls Fail to Fulfill Promise of Revitalizing Downtown" (Grossman 1987). Of course, it probably was unrealistic to expect the conversion of a single street to overcome decades of economic, social, and spatial trends which contributed to the decline of American downtowns. Nevertheless, the dissatisfaction with pedestrian mall performance has resulted in very few new pedestrian malls being constructed during the 1980s and 1990s, and some cities, such as Providence, Norfolk, Eugene, and New London (Connecticut) deciding to convert their pedestrian-only corridors back to streets allowing motor vehicles (see Houstoun 1990). At this time, it is still too early to tell if these conversions have increased economic activity on mall streets.

It is safe to conclude that American pedestrian malls, despite some success stories, have not fared as well as their European counterparts. To be fair, however, it must be pointed out that European cities possess several favorable conditions that make pedestrian street success more likely. First, the city centres of European cities have declined at a much slower pace than in America, thus providing a healthier environment for malls to prosper. Second, the combination of narrow medieval streets and old historic city centres make European cities easier to pedestrianize than American cities characterized by wider streets and newer buildings. Third, American downtowns usually lack the high density residential neighborhoods that characterize many European downtowns, thereby providing a supply of pedestrians throughout the day. Fourth, the greater use of privately-owned motor vehicles has made Americans more mobile than Europeans, and thus less likely to need and value pedestrian spaces. And fifth, the European culture seems to encourage the importance of streetlife and the recreational value of strolling to a higher degree than holds true in the United States (Rotoff 1981).

Table 2.1

Year of mall completion for
Brambilla and Longo's 1977 catologue

1959-1965	1966-1970	1971-1976	
Kalamazoo MI	Riverside CA	Eugene OR	Spartanburg SC
Miami Beach FL	Burbank CA	Evansville IN	Tampa FL
Knoxville TN	Danville IL	Parsons KS	Trenton NJ
Pomona CA	Minneapolis MN	Springfield IL	Vancouver BC
Fresno CA	Ottawa ON	Galveston TX	Wilmington DE
Jackson MI	Freeport IL	Las Cruces NM	Winchester VA
Atchison KS	Michigan City IN	Redding CA	Youngstown OH
Dallas TX	Salisbury MD	Richmond IN	Quebec City QB
Providence RI	Coos Bay OR	Lansing MI	Battle Creek MI
Santa Monica CA	Honolulu HA	Louisville KY	Greenville SC
	Oxnard CA	Monroe NC	Ithaca NY
	Pittsburgh PA	New London CT	Muncie IN
	Sacramento CA	Seattle WA	Paterson NJ
	Centralia IL	Toccoa GA	Portland ME
	Decatur IL	Baltimore MD	Reading PA
	E. Lansing MI	Frankfort KY	Rockford IL
	Lake Charles LA	Helena MT	Baltimore MD
	Lebanon NH	Napa CA	
		New Bedford MA	
		Oak Park IL	
		Tacoma WA	
		Atlantic City NJ	
		Poughkeepsie NY	
		Winston-Salem NC	
		Philadelphia PA	
N = 10	N = 18	N = 42	

Note: The sample encompassed pedestrian malls in 31 states and three Canadian provinces, thereby demonstrating the geographic diversity of this planning strategy. The most represented states were California (9 malls), Illinois (8), and Michigan (5).

Source: Brambilla and Longo (1977), pp. 174-175

Table 2.2

Year of mall completion for Rubenstein's 1978 catologue

1959-1965	1966-1970	1971-1977	
Fresno CA	Burbank CA	Napa CA	Auburn NY
Pomona CA	Oxnard CA	Redding CA	Ithaca NY
Santa Monica CA	Riverside CA	Redlands CA	Poughkeepsie NY
Miami Beach FL	Sacramento CA	New London CT	Greenville NC
Atchison KS	Honolulu HA	Wilmington DE	Monroe NC
Jackson MI	Centralia IL	Washington DC	Raleigh NC
Kalamazoo MI	Danville IL	Tampa FL	Winston-Salem NC
Providence RI	Decatur IL	Toccoa GA	Middletown OH
Knoxville TN	Freeport IL	Rockford IL	Painsville OH
Dallas TX	Michigan City IN	Springfield IL	Youngstown OH
	Lake Charles LA	Evansville IN	Eugene OR
	Salisbury MD	Richmond IN	Portland OR
	Minneapolis MN	Burlington IA	Allentown PA
	Winona MN	Dubuque IA	Erie PA
	Lincoln NB	Kansas City KS	Philadelphia PA
	Lebanon NH	Parsons KS	Philadelphia PA
	Coos Bay OR	Frankfort KY	Pittsburgh PA
	Galvaston TX	Louisville KY	Pottstown PA
	Ottawa ON	Portland ME	Reading PA
		Baltimore MD	Scranton PA
		Baltimore MD	Wilkes Barre PA
		New Bedford MA	Williamsport PA
		Newburyport MA	Greenville SC
		Salem MA	Spartanburg SC
		Battle Creek MI	Memphis TN
		E. Lansing MI	Dallas TX
		Lansing MI	Charlottesville VA
		Helena MT	Seattle WA
		Atlantic City NJ	Tacoma WA
		Cape May NJ	Sheboygan WI
		Paterson NJ	Quebec City QB
		Trenton NJ	Vancouver BC
		Las Cruces NM	
N = 10	N = 19	N = 65	

Notes: Thirty five states and three Canadian provinces are represented in the sample, with California (10 malls), Pennsylvania (10), and Illinois (6) being the most represented states. The regional breakdown is as follows: East (29 malls), Midwest (26), South (15), and West (21).

Source: Rubenstein (1978), pp. 180-185.

3 Case studies of American pedestrian malls

Confronted with the continuing exodus of retail sales from downtown to the suburbs, pedestrian malls became the symbol of hope to many city planners and downtown officials. In nearly 200 American cities, planners chose this strategy to help save the cores of their cities between 1959, when the first downtown mall was implemented in Kalamazoo (Michigan), and the mid-1970s. Recently, however, the image of these malls has become tarnished as evidenced by a headline in the June 17, 1987 issue of the Wall Street Journal: "City Pedestrian Malls Fail to Fulfill Promise of Revitalizing Downtown" (Grossman 1987). Very few new pedestrian malls were constructed during the 1980s, and many existing malls have been dismantled or converted from pedestrian-only streets to shared malls.

This chapter presents six case studies of pedestrian malls in American downtowns. The case studies, the research for which was conducted in the summer of 1988, include two traditional pedestrian streets (St. Cloud and Providence), one shared mall (Santa Cruz) and three transit malls (Minneapolis, Madison, and Denver). Table 3.1 provides background information on the six case studies. In terms of geographic distribution, each of the four time zones is represented. Financial limitations necessitated that the central region be overrepresented. Regrettably, the South was not included, although several southern cities (i.e., Charlottesville,

Memphis) were strong candidates for the sample. Following the case studies, the chapter will conclude by discussing 1) the extent to which downtown pedestrian malls have changed over time in terms of priorities, design, and economic activities, and 2) the key issues that planners must address for malls to remain viable into the 1990s.

St. Cloud, Minnesota: Mall Germain

St. Cloud is a growing small city (population = 42,566) located about 70 miles northwest of Minneapolis on the Mississippi River. The impetus for considering the construction of a downtown mall was a rapidly developing retail strip along Division Street on the edge of town that included an extremely popular enclosed shopping center called Crossroads. Originally, the primary goal sought with the newly conceived Mall Germain was to revitalize retailing in downtown St. Cloud. This proposed pedestrian mall was a controversial issue with the public from the beginning. The proposal was presented in a referendum that passed by the slimmest of margins (50.7% approved).

Mall Germain was opened in 1972. Only one street intersects the three-block long mall, and a walkway across this street is provided for pedestrians, clearly giving them the right-of-way. The mall is attractively landscaped and well maintained by the city parks department. It is anchored by a recently expanded department store--Herbergers--and is situated in a downtown that has been the scene of several large-scale developments in recent years.

The commitment by the owners of Herbergers in 1985 to remain and expand was a boost for downtown retailing. However, other indicators suggest problems for the retail sector. The percentage of mall establishments engaged in the sale of retail goods (53.3%) was the lowest of the six malls studied (see Table 3.2). The 14.3% vacancy rate for mall establishments was the second highest of the sample. Also, the retail mix is questionable, seeming to be somewhat outdated and not accurately reflecting the increasing number of young white-collar office workers in the area or the 16,000 students who attend St. Cloud State University, which is eight blocks from Mall Germain.

There is no mechanism to control or monitor design

on the mall. Thus, as one would expect, some noteworthy design-related problems exist. The mall is lined with turn-of-the-century Victorian buildings with facades that have been altered considerably over the years, resulting in a wall along the street that is chaotic-looking and unattractive. Herbergers is a suburban-style building with no street-level windows on the mall. This block-long blank wall contradicts the very essence of a pedestrian space, repressing street activity (see Figure 3.1). Finally, some believe that the mall is one block too short. If it were extended one more block to the east, as was originally planned, it would connect the retail core with the new convention center and the river.

The most obvious shortcoming of the mall is that it is empty of people most of the time. Rarely does one see the mall bustling with pedestrians. The lack of people-attracting features (such as street vendors, outdoor cafes, and street performers) contributes to this inactive state. Currently, these types of activities are discouraged; in some cases, they are not permitted by ordinance.

Despite these problems, public opinion regarding Mall Germain is not entirely negative. Although most would agree that the mall has failed to fulfill original expectations, they also concede that the downtown might be in worse condition had it never been built. Moreover, many believe that the mall has contributed to the aesthetic and open space quality of the downtown area. In many ways, St. Cloud's Mall Germain is representative of the current status of a majority of traditional pedestrian streets constructed in the 1960s and 1970s.

Providence, Rhode Island: Westminster Mall

In the late 1950s, planners for Rhode Island's capital city (population = 156,804) introduced the concept of a pedestrian mall as a part of the Downtown Providence 1970 plan. The purpose was to stimulate downtown retailing and to prevent the exodus of shoppers to the suburbs. The idea had the support of the majority of the downtown business community.

Westminster Mall is a cross-shaped mall consisting of 4 blocks of Westminster Street and 1 1/2 blocks of Union Street, which intersects with Westminster near the center of the mall. The mall is enhanced by a large collection of handsome commercial buildings dating from the late nineteenth century. Vehicular traffic is permitted to cross the mall at two locations; however, there is only one lane of one-way traffic, and vehicles must stop for pedestrians. Pedestrians clearly have first priority on the mall (see Figure 3.2).

Retail activity has declined steadily, partly as a result of the intense competition from 12 major shopping complexes in suburban Providence. Some of the traditional downtown stores (such as Woolworths and several drug stores) have persevered, and a few chain stores were opened on the mall during the 1980s. But only one department store (Peerless) remains on the mall that housed four in the 1960s. Vacant space also is a major problem. In 1988, 16.4% of the mall's ground-floor establishments were vacant--by far the highest vacancy rate of the six malls studied (see Table 3.3). Moreover, a 1981 study showed that the vacancy rate for total available floor space for Westminster Mall was twice that of its two parallel streets, Washington and Weybosset (Allen 1981).

If one were to venture onto the mall on a typical weekday between 11:00 a.m. and 3:00 p.m., one would be impressed with the abundance of activity taking place. Most of this activity is generated by office workers from the nearby financial district who are shopping, eating, and strolling on their lunch break. At other times during the day, however, activity is minimal; after 5:00 p.m., virtually all activity has ceased. One clue to the activity patterns is that street vendors pack their wares and head home between 2:00 p.m. and 3:00 p.m. The mall has no restaurants or nightclubs to stimulate activity in the evening.

Other problems stem from the appearance of Westminster Mall. It was the most poorly maintained mall of the sample; it contained far too much trash and litter (especially on weekends), planters with no plants or flowers, and vandalized lights and signs that had not been repaired for quite some time. In terms of design, a number of people think that the modern-looking landscape on the street gives the mall a cluttered look, and does not complement the nineteenth century buildings lining the street. Finally, a utilitarian police booth in the middle of the mall not only adds nothing to the area's aesthetics, but creates the impression that the mall is unsafe and that police surveillance is needed.

Westminster's immediate future is set: city officials have decided to transform it into a shared mall in which limited automobile traffic is permitted. The plan, which is supported by the mayor, city planning department, and local newspapers, is to close the mall to vehicles between 11:00 a.m. and 2:00 p.m. and to permit parking and one lane of traffic the rest of the time. Downtown merchants are split on the merits of this plan. Planners foresee a new purpose for the mall in the 1990s: the new focus will be on housing, entertainment, and service retailing, the latter oriented towards office workers.

Santa Cruz, California: Pacific Garden Mall

The small city of Santa Cruz (population = 41,483) is located south of San Francisco on the Monterey Peninsula. The reasons for constructing a mall were based not so much on the threat of suburban competition as on concerns about the physically and economically deteriorating downtown and on the desire to improve the city's social environment. A progressive city council sought a planning strategy that could incorporate a more attractive and prosperous downtown retail center with a park-like community gathering place. The answer was the Pacific Garden Mall, a shared mall that opened in 1969.

The mall encompasses six blocks of Pacific Avenue in the heart of downtown. Unlike in most cities, the traffic patterns were not seriously disrupted with mall implementation because no streets intersect this segment of Pacific Avenue; rather, all the streets terminate into the avenue.

Automobiles have been allowed on the mall from the beginning. However, despite the presence of motor vehicles, the primary orientation is toward the pedestrian. Drivers are required to stop for pedestrians along the six blocks, and because there are no cross streets in this area, the problem of pedestrians having to wait for traffic is avoided (see Figure 3.3). Local officials believe that allowing vehicular access to the mall makes it easier for people to become familiar with mall establishments. In addition, diagonal parking is provided in some sections of the mall. There was disagreement about the merits of on-mall parking. Some thought that, despite the paucity of spaces provided, it was worthwhile to give the "illusion of available parking"; others said that the parked cars detracted from the appearance of the mall and that precious space that could be better utilized for pedestrian needs was wasted.

The overall appearance of Pacific Garden Mall creates an extremely positive impression. It has been attractively landscaped to look like a botanical garden, with many varieties of trees, shrubs, and plants. The mild California climate has helped to produce a lush downtown corridor. Ironically, some have complained that the vegetation is so lush that sight lines to storefronts are blocked, and dark potentially dangerous places are created. The mall also appeared as the cleanest and best-maintained of the six malls visited. A lasting visual image is of shopkeepers sweeping in front of their stores prior to opening in the morning.

Retailing along the mall was conducted primarily in small business establishments owned by independent entrepreneurs who tended to market their goods and services to tourists from nearby beaches as much as to local residents. The lack of chain stores was especially noticeable. Compared to the other malls studied (see Tables 3.2 and 3.3), the Pacific Garden Mall in Santa Cruz had the second-highest number of total mall establishments (106), the second-highest percentage of establishments dealing in retail goods (67%), and the lowest ground-floor vacancy rate (0.9%).

Perhaps the major concern about the mall today is social in nature. Many are worried about the increasing numbers of "street people" who inhabit the mall and intimidate would-be shoppers. Included in this group are the homeless, public inebriates, drug dealers and users, teenage punk rockers, and such transients as "deadheads" from San Francisco. Several of the interviewees suggested that the original objective of having the mall function as a social gathering place had been overachieved and that it was time to redesign the mall to operate more as a shopping center.

The consensus was that Pacific Garden Mall has had a positive impact on Santa Cruz because it served to preserve the architecture and pedestrian quality of downtown and because it allowed the downtown to maintain its village-like ambiance. However, because the mall is 20 years old, most interviewees thought that the time was right for reassessment and updating. Changes in design were discussed most frequently, such as the removal of some trees, shrubs and parking places, the addition of sidewalk cafes, and modifications in landscaping and use of certain spaces. Among other reasons, interviewees thought that the implementation of some of these suggestions would help eliminate the "dead spots" where the street people congregate.

In October of 1989, one year after the research for this case study was conducted, a devastating earthquake hit downtown Santa Cruz. Much of the Pacific Garden Mall lay in ruin. More than one-third of the buildings lining the mall, most of them historic turn-of-the-century structures, were damaged beyond repair (Pepper 1989). Large tents were erected nearby the mall to provide a temporary place for mall merchants to sell their goods during 1989's Christmas season. The city is presently in the midst of an ambitious rebuilding effort designed to restore and, where possible, improve the Pacific Garden Mall.

Minneapolis, Minnesota: Nicollet Mall

Retail merchants located in downtown Minneapolis (population = 370,951) faced the competition from suburban shopping malls earlier than those in most cities as a result of the opening in 1956 of Southdale, reported to be the first enclosed suburban mall in the nation. When Nicollet Mall was dedicated in 1967, it was the centerpiece of downtown Minneapolis and the prototype transit mall for America. Planners, architects, and city officials came from around the world to view this innovative example of downtown planning.

Originally the mall included eight blocks along Nicollet Avenue; in 1982, it was extended for two additional blocks. It was the first transit mall to be constructed in the United States. It combined a 24-foot-wide serpentine bus lane that was flanked by pedestrian walks of up to 36 feet wide. Among the amenities were attractive bus shelters and a heated sidewalk, both useful features during the harsh Minnesota winters. When it first opened, the mall was lined with small shops and anchored by numerous department stores, both of which were oriented toward pedestrians on the street.

The buses have always been an important feature of

the mall. On an average weekday over 30,000 riders use the six routes that traverse Nicollet Mall. The buses provide downtown workers and shoppers with convenient access not possible by automobile. The diesel-powered buses have been criticized, however, because their noise and exhaust discourages the presence of pedestrians, sidewalk cafes, and other street-level activities on the mall.

Retailing on Nicollet Mall has undergone a significant metamorphosis. Until the mid-1980s, five department stores were situated on the mall. Today, three of the original five have closed there doors, but have been replaced by two new upscale department stores: Saks 5th Avenue and Neiman Marcus. The small independent shops that once predominated on the mall have been replaced by chains that are located off indoor courts in large shopping complexes. In 1977, for example, 72 retail/service establishments were located directly on the mall between Washington and 10th Streets; by 1987 that number had been reduced to only 31. These changes have caused some concern that the retail mix is becoming too upscale and that the mall is not providing sufficient shopping opportunities for a wide cross section of people.

The level and intensity of street activity has been declining as well. The reduction of street-level shops is both a cause and a result of this trend. Another factor affecting street activity is the increasingly popular and much- heralded skyway system that connects over 35 contiguous downtown blocks. The skyway system is in competition with the pedestrian system on the mall because the two are poorly integrated (see chapter 6). Because of access to the skyways and the proliferation of enclosed downtown shopping complexes, people choose to stay indoors during all seasons, much to the detriment of Nicollet Mall.

The mall also is burdened with far too many dead spaces that inhibit human activity. These dead spaces are the result of a variety of factors, ranging from a disastrous 1982 fire to the presence of institutional facilities (such as the Federal Reserve Bank) that cause gaps in retail activity to the increasing presence of blank, uninteresting walls fronting the mall. As with many downtown malls, little street activity occurs during the evening on Nicollet. The percentage of mall establishments serving food and drinks (13%) was the lowest of the cities examined (see Table 3.2), and there are no nightclubs or similar establishments on the mall.

For a frostbelt city, Minneapolis is indeed fortunate to possess such a prosperous downtown. The key question is: to what extent will Nicollet Mall continue to contribute to the prosperity? To help address that question, the Nicollet Mall Implementation Board, consisting of representatives from city hall and the downtown business community, was established in 1987. Among other things, the board recommended that the mall be redesigned, a recommendation that was accepted by the city and completed during the early 1990s. The future prosperity of Nicollet Mall may well depend on the success of the huge Mall of America - the largest shopping mall in the world located in suburban Bloomington - which opened in 1992.

Madison, Wisconsin: State Street Mall

The 1960s was a turbulent time for the downtown of Wisconsin's capital city (population = 170,616). Anti-war demonstrations and student protests had caused serious damage to the retail district, in terms of both reputation and vandalized buildings. Many people were choosing to avoid the downtown. To counter this trend, planners devised the State Street Mall, a six-block transit mall on the traditional main shopping street. Nicollet Mall in Minneapolis served as a primary model for the mall's designers; even the serpentine bus lane was included.

State Street Mall was constructed in several phases. The first phase opened in 1977, and the mall was not completed until 1982. Two tremendous activity generators anchor the mall. At one end is the Wisconsin State Capitol, housing nearly 25,000 government workers. The domed capitol is a marvelous focal point; it is visible from any point along the entire length of the mall. At the other end is the campus of the University of Wisconsin with 44,000 students and 13,000 employees. The mall, lined with older buildings containing small and colorful shops, connects these two anchors that are sources of steady pedestrian traffic for the street.

Several indicators confirm that this was an economically successful planning strategy. A recent study (Downtown Madison Partners Inc. 1986) showed that the average base rents for commercial space on the mall was $9.87 per square foot, which is considerably higher than the downtown average of $8.15 per square foot. The vacancy rate (3.4%) for ground-level mall establishments was the second lowest in the sample of six cities (see Table 3.3). Finally, the fact that State Street Mall contained far more mall establishments than were contained in the malls of the other five cities in the study indicates not only the prime location but also the predomination of small shops (see Table 3.2). Of note, there are no department stores on State Street or in downtown Madison.

The retail mix is what one might expect for a college town: record shops, bookstores, poster shops, coffee shops, clothing stores catering to the young, and ethnic restaurants. The recent in-migration of chain stores and fast-food restaurants is causing concern that the independent storeowner's sales are being cut sharply and their rental fees are being increased. State Street Mall offers the most active nightlife of any of the malls studied. It includes a movie theater, a concert hall, and many student-oriented bars and restaurants. The mall contains a higher percentage of establishments serving food and drinks (28%) than any of the other malls (see Table 3.2).

Buses are an important feature of the mall, dominating the street scene to a greater extent than in Minneapolis. More than 700 buses along 16 routes traverse State Street every weekday. The mall has several attractively designed bus shelters for passenger comfort. As in Minneapolis, there have been numerous complaints about the noise and pollution the buses produce. Moreover, some people contend that the amount of space devoted to the bus lane hinders certain activities on the mall (for example, outdoor cafes, community events).

Of the six malls examined, State Street appears to be the least oriented towards pedestrians. The relatively narrow sidewalks seem dwarfed by the massive bus lane, leaving little room for pedestrians or activities. Moreover, pedestrians have to cross busy streets at four locations when walking the length of the mall. Generally, these wide busy streets can be crossed only with a green light signal, and the time between lights is annoyingly long. Pedestrians also have to be on the constant lookout for a tremendous number of bicyclists who share the mall with buses and pedestrians.

Measured in economic terms, Madison's mall must be proclaimed a success. However, there is considerable disagreement about the appropriate mix of transportation. Some would prefer that the part of the mall closest to the capitol be restricted for pedestrians use only. Others advocate opening the mall to limited automobile use and perhaps some parking. The Madison Metro Transit System maintains that State Street must remain open to their buses for the entire length of the mall. Given the success of this mall, it is unlikely that any major changes will take place in the near future.

Denver, Colorado: The 16th Street Mall

Very few pedestrian malls have been built in the 1980s. One exception is the widely publicized 16th Street Mall in Denver (population = 492,365). The impetus for establishing the mall was a declining downtown retail sector combined with a need to improve downtown and areawide transit service. The idea for a transit mall was proposed jointly in the late 1970s by Downtown Denver, Inc. (the downtown association) and the Regional Transportation District (RTD), the transit authority. From the beginning transit was a crucial factor in the mall's evolution, as evidenced by the fact that the Urban Mass Transit Administration funded 80% of the mall's construction costs.

The transit mall opened in 1982. At first glance, its length of approximately 1 mile (13 blocks) appears to be too long and spread out to sustain sufficient density levels. However, the mall does not seem overly long when walking it; the design features and layout contribute to an interesting and enjoyable walk. The 80-foot-wide right-of-way is centered by a 22-foot-wide walkway flanked by two 10-foot-wide bus lanes, with a 19-foot-wide sidewalk on each side (see Figure 3.4).. The mall features two excellent focal points: On the eastern end is the Colorado State Capitol, and the D & F Tower is a highly visible 1930s landmark on the western end. Authorities estimate that about 90,000 pedestrians use the mall each day.

The 16th Street Mall's most apparent and innovative feature is the shuttle bus system that traverses the length of the mall. Transit terminals at each end of the mall receive the majority of bus lines serving Denver and its suburbs; these two terminals were opened in 1983 and 1984. Electrically powered shuttle buses connect the two terminals along the mall. These quiet shuttles stop on every block and are free of charge to anyone wishing to ride. At rush hour, the time between shuttles is only 70 seconds. They have proven to be so popular that on an average weekday over 40,000 people ride the shuttle; about one-half the riders transfer from other buses (see Figure 3.5).

Retailing on 16th Street has turned around since the mall's completion. The mall has a higher percentage of establishments engaged in the sales of retail goods (69%) than any of the other malls, and its ground-level vacancy rate (4.6%) is lower than the sample average (see Tables 3.2 and 3.3). The majority of the stores have their main entrances on the mall, although there are a few indoor minimalls and one sunken plaza. Even the successful Tabor Center--a large enclosed shopping center on the western end--has been carefully designed so to accommodate pedestrians on the street. Tabor Center's facade is transparent, so shoppers inside can see the pedestrian mall clearly and pedestrians on the 16th Street mall can easily see what stores are inside. A group of vending carts with tempting wares just outside the center helps to integrate the two shopping areas. One frequently voiced complaint was that the stores close too early in the evening.

A number of strategies have been employed to help maintain the mall's vitality. Street vendors are encouraged to sell food, crafts, souvenirs, and the like, on the mall to produce more activity. Restaurant owners have been encouraged to provide outdoor seating. Movable chairs have been provided by the city and placed throughout the mall that allow people to sit where and with whom they want. Attractive lighting and horse-drawn carriage rides help to stimulate activity in the evening.

Although problems exist with the mall, they generally are less serious than those with other malls. Many believe that there is a need for higher-quality use of the open space adjacent to the mall. Several large vacant spaces, currently used as surface parking lots, cause great concern for both economic and aesthetic reasons; these spaces were referred to as "gaps in the smile" by one of the individuals interviewed. Another concern is about the extremely wide and busy cross streets; pedestrians frequently are interrupted in their journeys to wait for the heavy volume of traffic to pass.

Today, the 16th Street Mall must be declared the state-of-the-art American pedestrian mall. Planners, public officials, and urban scholars flock to it just as they did to Minneapolis's Nicollet Mall in the past. One of the keys to its success is the level of private/public management that oversees mall activities. A tremendous amount of attention is given to mall programming and promotion (see Zehnpfenning 1988). Denver's mall is clearly the model of success for the 1980s at which planners in other cities are looking for guidance.

Shifting trends for America's pedestrian malls

More than 30 years have passed since the first pedestrian mall was constructed in the United States, and the fifteenth-year anniversaries of most American malls already have been celebrated. To what extent have downtown malls changed during this period in terms of design, economic activities, and issues?

The most fundamental shift is the distribution of transportation on the malls. Most of the malls that were constructed in the 1960s and early 1970s were traditional pedestrian streets. On these malls, transit and motor vehicles were not permitted, and the pedestrian truly reigned as "king of the road". However, in many of these cities, pedestrian traffic was found to be insufficient to maintain consistently high levels of street activity.

To rectify the situation, city planners in an increasing number of cities are considering opening their malls to transit systems or limited automobile traffic. The prevailing theory is that buses or cars will provide more steady levels of mall activity and the reduced space for pedestrians will result in greater densities of people on the mall. The case studies clearly support this trend. In terms of economics and overall mall vitality, the most successful malls in this study are the three transit malls and the shared mall. The two malls in which vehicles are not allowed--Providence and St. Cloud--exhibited vacancy rates of at least 2 1/2 times the sample average (see Table 3.3). Moreover, a commitment already has been made in Providence to convert Westminster Mall into a shared mall, and discussions on the idea have begun in St. Cloud.

A second change has been the increasing orientation toward indoor activity. Previously, virtually all mall establishments were oriented toward the mall street traffic, thereby directly contributing to the numbers of people walking on the mall. A trend of the 1980s was to construct indoor shopping centers establishments fronting interior courts. If the enclosed plaza is large enough, many people will seldom venture out onto the mall street, thereby limiting street activity and the viability of street-level establishments. The proliferation of such enclosed centers already has occurred in large cities like Minneapolis and Denver, and even is an issue in a small city like St. Cloud.

This orientation towards indoor activity is intensified when the alternative of a weather-protected pedestrian system exists. The classic example is the skyway system in Minneapolis, which, as mentioned earlier, connects over 35 contiguous downtown blocks, and crosses Nicollet Mall in several places. Because no direct connections exist between the two pedestrian systems, they indeed are competitive, not complementary. The enormous popularity of the skyways has been a major factor in the decline of Nicollet Mall.

A significant change in retailing is taking place: Chain stores and restaurants gradually are replacing the independent locally owned enterprises that once dominated malls. This trend is particularly evident in the larger cities and within enclosed shopping complexes. In Madison and Santa Cruz, the potential existence of chains along their malls is cause for concern. People are worried that the influx of chains will lead to increased rents, thus forcing some owners of local establishments to close or relocate their stores, and that the owners of chains will not exhibit the level of downtown commitment and civic pride that is characteristic of many of the independents owners.

A final change involves the increased attention being devoted to the management of the malls. Most of the attention given to the early malls focused on design and occurred during the planning stages. A mistaken assumption made by many city planners was that people would automatically flock to the malls once they were built. Today, they have learned that coordinated management of malls is essential to success. Private/public partnerships, as exemplified by the arrangement between the City of Denver, the transit authority, and the Denver

Partnership, are being used effectively to manage the promotion, maintenance, and programming of malls. This increased emphasis on management is best summed up by the attitude assumed by Denver officials who realize that "The mall was a tool for revitalization, not an end product of revitalization" (Boulter 1988, p. 4).

Key issues for the 1990s

The changes described in the previous section can be translated as some of the key issues for pedestrian malls in the 1990s. Nationwide, downtown planners are grappling with issues concerning the balancing of pedestrian and vehicular needs, the integration of indoor-oriented and street-level activities, and the coexistence of national chain stores and local independent entrepreneurs.

A number of other important issues must be be addressed as well. One concern is the increasing number of street people frequenting the malls. In Providence, Santa Cruz, and Madison, the presence of the homeless, public inebriates, punk rockers, and/or drug users on the downtown malls is thought to inhibit shoppers from using the mall establishments. Public safety and the values of suburban middle-class shoppers are the key factors underscoring this concern. However, on the other side of this issue is a civil liberties argument that malls are public places in which everyone has an equal right to spend time and therefore exclusive policies are both unconstitutional and immoral (see Figure 3.6).

A second issue is the lack of evening activity. In the typical scenario, one would find the malls to be fairly active during the day, with a retail sector geared toward daytime office workers. However, few activities are offered after 6:00 p.m., and mall activity levels quickly drop. In most of the case studies, few stores remained open in the evening, and virtually no forms of nighttime entertainment (such as bars or nightclubs) existed. The State Street Mall in Madison is a notable exception to this trend, but the nearby presence of a large university with a young student population that is attracted to the mall probably explains this anomaly. How to attract nighttime activity to the mall is a difficult issue for planners.

A third issue that must be addressed is the

increasing amount of dead space that exists on pedestrian malls. Malls originally were designed to contain a continuous wall of activity in the form of shops and services oriented toward pedestrians on the street. Today, huge gaps exist in this wall on many malls. Some of these gaps were unplanned, resulting from failed businesses or demolished buildings. Other dead spaces result from the erection of new buildings that were not designed to integrate effectively with the street level. Buildings with blank, uninteresting walls fronting the street are prevalent in downtowns across America and are especially devastating to pedestrian spaces. Successful pedestrian spaces require an uninterrupted sequence of activity in order to sustain a vital urban district.

Other problems will continue to be of concern during the 1990s, such as the noise and pollution caused by diesel buses on transit malls and the constant interruption of pedestrians who must negotiate busy cross streets. However, these problems are relatively minor compared to the most fundamental factor affecting America's pedestrian malls: the decline in the importance of walking in America. In our automobile-oriented culture, a low priority is placed on walking, and consequently people are not all that concerned with the quality of their pedestrian spaces. Unlike in European countries, where strolling along city streets designed for pedestrians could be labeled as a national pastime, most Americans prefer not to walk. When they do choose to walk, the increasing preference is to do so in an enclosed climate-controlled environment.

This is not to suggest that pedestrian malls cannot have a place in the future fabric of American downtowns. Without malls and other pedestrian-oriented spaces, the urban environment would fulfill even fewer human needs than it does today. Success stories in cities large (Denver) and small (Santa Cruz) demonstrate that malls can contribute to downtown revitalization. What downtown planners must understand is that, given current cultural values, the majority of American cities will be unable to maintain pedestrian-only malls and therefore, in many cases, city planners will have to design malls that will accommodate the needs of a motorized public along with those of the pedestrian.

Table 3.1: Background Information on American Case Studies

City	1990 City Population	Mall Name	Mall Type	Year Opened	Length
Denver, Colorado	467,610	16th Street Mall	Transit	1982	13 blocks
Madison, Wisconsin	191,292	State Street Mall	Transit	1977	6 blocks
Minneapolis, Minnesota	368,483	Nicollet Mall	Transit	1967	10 blocks
Providence, Rhode Island	160,728	Westminster Mall	Traditional*	1965	4 blocks by 1 1/2 blocks
St. Cloud, Minnesota	48.812	Mall Germain	Traditional	1972	3 blocks
Santa Cruz, California	49,040	Pacific Garden Mall	Shared	1969	6 blocks

* coverted to shared mall in 1989-1990

Table 3.2: Inventory of Mall Establishments - 1988

City	Total Establishments	Retail Goods	Business Services	Personal Services	Food/ Drink	Other
Denver CO	87	60 (69%)	7 (8.1%)	4 (4.6%)	15 (17.2%)	1 (0.1%)
Madison WI	143	83 (58%)	3 (2.1%)	11 (7.7%)	40 (28%)	6 (4.2%)
Minneapolis MN	69	37 (53.6%)	7 (10.1%)	6 (8.7%)	9 (13%)	10 (14.5%)
Providence RI	46	27 (58.7%)	6 (13%)	2 (4.3%)	9 (19.6%)	2 (4.3%)
St. Cloud MN	30	16 (53.3%)	2 (6.7%)	5 (16.7%)	5 (16.7%)	2 (6.7%)
Santa Cruz CA	106	71 (67%)	9 (8.5%)	5 (4.7%)	19 (17.9%)	2 (1.9%)
TOTALS	481	294 (61.1%)	34 (7.1%)	33 (6.9%)	97 (20.2%)	23 (4.8%)

Note: Inventory conducted during summer/fall of 1988.

a. Does not include vacancies (see Table 3.3).
b. Includes establishments such as theatres, hotels, goverment offices, churches, and private clubs.

Table 3.3: 1988 Mall Vacancy Rates

City	Total Units Available	Vacancies	Vacancy Rate
Denver CO	91	4	4.6%
Madison WI	148	5	3.4%
Minneapolis MN	74	5	6.8%
Providence RI	55	9	16.4%
St. Cloud MN	35	5	14.3%
Santa Cruz CA	107	1	0.9%
TOTAL	510	29	5.7%

Note: Number of vacancies based on inventory conducted during summer/fall of 1988.

a. Total Units Available includes all existing establishments plus all vacancies.
b. Surface parking lots are counted as vacancies.

Figure 3.1: St. Cloud Mall Germain. The facade of this newly constructed department store is faceless and uninviting and does not integrate effectively with Mall Germain (1989)

Figure 3.2: Providence Westminster Mall. Automobiles must stop when crossing Westminster Mall because pedestrians always have the right of way (1988)

Figure 3.3: Santa Cruz Pacific Garden Mall. Pacific Garden Mall is an example of a shared mall, one in which limited automobile traffic is permitted (1988)

Figure 3.4: Denver 16th St. Mall. The design of 16th Street Mall includes two bus lanes and ample pedestrian space (1988)

Figure 3.5: Denver shuttlebus. One of the many electronically powered shuttle buses serving shoppers on the 16th street Mall (1988)

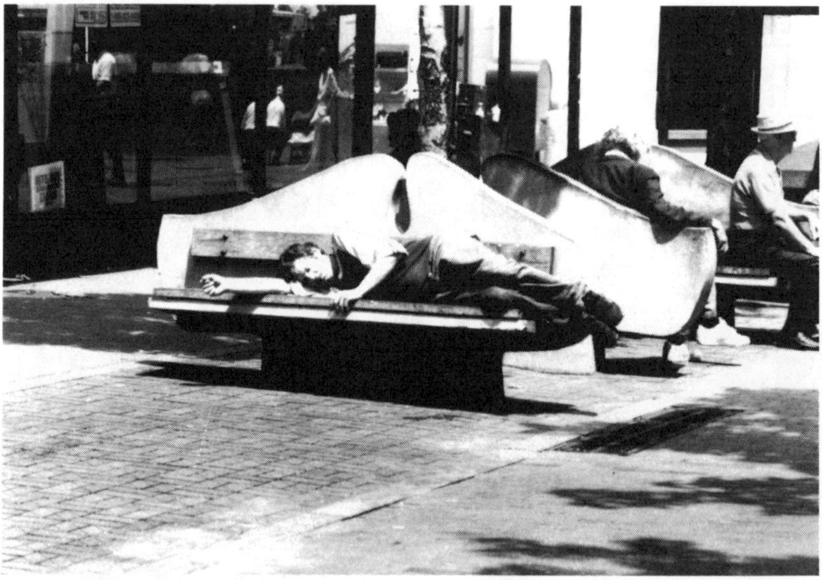

Figure 3.6: Providence streetpeople. The presence of street people is intimidating to mall merchants and middle-class shoppers (1988)

4 Pedestrian streets in Sweden's city centres

City centres in Sweden were threatened by a new enemy starting in the 1950s: The private automobile. The increasing use of the automobile was clearly incompatible with the cultural, economic, and physical fabric of city centres, most of which were designed to accommodate the efficient movement of pedestrians and mass transit. Consequently, the narrow streets were clogged with vehicular traffic, noise and air pollution approached critical levels, the conservation of historic buildings became increasingly difficult, walking was becoming a more dangerous activity, and a "way of life" characteristic to city centres for generations was in great peril.

In response to this problem, most Swedish cities prepared traffic management plans during the 1960s and 1970s which, among other strategies, recommended that pedestrian-only streets be created within the commercial core of the city centre. The resulting pedestrian streets came in many forms, but most commonly were located along a central corridor--usually a few linear blocks along the traditional main shopping street--where pedestrian transportation was given the highest priority. In 1961, Lund, Stockholm, Gothenburg, and Helsingborg were among the first cities in Sweden to restrict automobiles from selected central streets, and by the end of the decade, 35 cities possessed pedestrian streets (Gunnarsson and Lindstrom 1969). Today, experts estimate that over 100 Swedish cities have pedestrian streets in their city centres.

Over the years, pedestrian streets have been accepted as an integral component of city centres. However, changes in shopping and transportation preferences amongst the Swedish people are forcing city planners to reassess the role, design, and retail mix of pedestrian streets. Increased competition from suburban shopping areas is effecting sales along pedestrian malls, and therefore is of tremendous concern to the downtown business community. Thus, a number of cities are contemplating changes for their pedestrian streets during the 1990s.

The purpose of this chapter is to examine the city centres of six Swedish cities to help us better understand 1) the extent to which pedestrian streets have changed over time in terms of design, activity patterns, transportation factors, and retail trends, and 2) the key issues that city planners must address for pedestrian malls to contribute to city centre vitality during the 1990s. In addition, this analysis will permit a direct comparison between malls in the United States with those in a European country, which is presented in the subsequent chapter. The six pedestrian street systems to be studied are located in the cities of Falun, Gothenburg, Linkoping, Orebro, Stockholm, and Uppsala (see Figure 4.1). One of the primary reasons Sweden was selected for study is that despite the widespread application of pedestrian streets in Swedish cities, little has been written about this topic with the exception of Nordquist's (1984) Swedish language work.

Brief descriptions of the case studies

This section will provide capsule overviews of each of the six case studies by briefly describing the characteristics and features of their pedestrian streets.

Falun

The smallest and northernmost city in the sample is Falun (population = 52,448). The city centre contains three pedestrian streets, all of which terminate into the spacious Storatorget (Great Square). By far the most active street is Hotmgatan, which houses 3 department stores and the majority of fashionable chain stores. In contrast, Asgatan contains mostly secondary shops and services, while Staaggatan is dominated by government and office buildings. A busy street for buses intersects the pedestrian mall by the big square. Several indoor shopping complexes have opened recently on the mall. The streets were fairly active, although to a lesser extent than the other five case studies.

Gothenburg

Known as Goteborg to the Swedes, the country's second largest city (population = 430,763) is Sweden's major port. The core of its system of walking streets is Kungsgatan, a long street lined with busy shops and restaurants. Intersecting Kungsgatan are Fredsgatan, a narrow shopping street housing two major department stores, and Korsgatan, a less active street with small shops next to the cathedral. Fredsgatan also leads to Nordstan, Sweden's largest indoor shopping center with over 130 stores. A number of busy streets cross Kungsgatan, including two wide thoroughfares for automobiles and trams. Activity levels, especially on Fredsgatan and the eastern portion of Kungsgatan, were consistently high. There were also a large number of restaurants, many with outdoor seating directly on the mall.

Linkoping

The city centre of Linkoping (population = 119,167) is the most pedestrian-oriented of the sample. The pedestrian system, which is linked with abundant open spaces and parks, extends beyond the commercial core to the cathedral district and into several adjacent residential neighborhoods. The commercial segment of the pedestrian street system focuses on three squares and their connecting streets. Tannesforsgatan is a picturesque, winding, medieval street lined with smaller shops that connects Storatorget with Tradgardstorget. The large department stores and chain establishments are situated on Repslagaregatan and Nygatan, both of which feed into Lillatorget (Little Square). A very busy bus street crosses Nygatan. The mall is very attractively designed with a fountain, covered walkway, trees and shrubs, and a generous amount of sitting space.

Orebro

The pedestrian system of Orebro (population = 119,824) consists of two parallel streets: Drottninggatan, a transit mall allowing buses, bicycles, and pedestrians, and Kopmangatan, a short pedestrian-only street constructed in 1982. These streets connect two major squares: Stortorget and Vaghustorget. Overall, Orebro's city centre is the least pedestrian-oriented of the cities visited, primarily because 1) pedestrians are overshadowed by buses and bicycles on Drottninggatan, 2) parking is permitted directly on Stortorget, and 3) pedestrian crossings approaching the mall are very dangerous. Major changes are planned in the near future, especially for Drottninggatan.

Stockholm

Sweden's largest city (population = 669,485) has a pedestrian system consisting of three primary streets, each with its own distinctive look and character. Vasterlanggatan, located in Old Town, is an extremely narrow and winding medieval street lined with small fashionable shops and cafes. Drottninggatan is the traditional shopping corridor that contains the large department and chain stores; a walking bridge over the water connects this street to Vasterlanggatan. Finally, Sergelgatan is a newly-created street that was introduced as part of a major urban renewal project in the 1960s (see Figure 4.2). The entire system is crowded with pedestrians, vendors, and performers throughout the day. The overall design scheme is simple and basic, with little attention given to landscaping or other design amenities.

Uppsala

The idea for placing pedestrian streets in Sweden's fourth largest city (population = 161,828) appeared in Uppsala's 1955 master plan, but due to a well-orchestrated resistance by the downtown business community, was not implemented until 1964. The city's commercial pedestrian system is contained primarily on a single linear corridor: Kungsangsgatan/Svartbacksgatan (see Figure 4.1). A large square (Storatorget), located at the center of this corridor, serves as the main bus terminal for the city. This provides excellent linkages between pedestrians and transit, but the volume of bus traffic tends to overwhelm pedestrians, especially near the square. Several indoor shopping complexes have been built on the street in recent years.

Transportation considerations

The initial factors underscoring the building of pedestrian streets in Sweden are related primarily to transportation. Unlike downtowns in Germany and the United States, which used pedestrian malls as an urban renewal strategy to combat city centre deterioration caused by war and suburbanization, respectively, Swedish cities constructed pedestrian streets to facilitate the movement of traffic. The influx of automobiles onto narrow downtown streets caused serious problems in terms of safety, congestion, and environmental quality.

In each of the six cities, planners realized that a coordinated traffic management plan that promoted the separation and linkage of different transportation systems was the answer. This concept, applied throughout Sweden, took into account the transportation needs of people utilizing walking, bicycles, public transit, automobiles, and railroads. The needs of the pedestrian were paramount to this comprehensive approach to planning, which sought to improve pedestrian safety and to provide a more human-scale environment. Consequently, the implementation of pedestrian streets where vehicular traffic was limited or restricted became a commonly applied planning strategy in city centres across Sweden. Pedestrian safety statistics provided by the Swedish Road and Traffic Research Institute in Linkoping and observations of pedestrian street activity levels suggest that these goals have been realized.

The interface between pedestrians and other forms of transportation is another key objective of comprehensive traffic management. Bicycles are used for commuting purposes by many Swedes. According to Pucher (1990, p. 447), a higher percentage (49%) of transport trips are made by pedestrians and bicyclists in Sweden than in any other European country. For example, it is estimated that over 15,000 people come into Orebro's city centre by bicycle on a typical summer day. Although bicycle riding is prohibited on most of the pedestrian streets, ample and convenient bicycle parking is often offered closeby. In Uppsala, facilities for parking one's bicycle is provided on the sidestreets that intersect the pedestrian mall, while in Falun and Orebro, bicycle racks can be found directly on the pedestrian streets. These racks were heavily used, attesting to the bicycle's popularity. At times, however, bicycles can be a nuisance and a safety problem to pedestrians; this was especially true in Orebro, where pedestrians need to watch out for bicyclists moreso than for motor vehicles.

Public transit is also widely utilized in Swedish cities. In Gothenburg, for example, about twice as many people enter the city centre on buses and trams as drive automobiles (Gothenburg Planning Department 1989). The linkage between pedestrian streets and transit was clearly evident in each city. The major downtown bus terminals in Uppsala, Falun, and Linkoping are situated on central squares where the pedestrian mall is intersected by a bus-only thoroughfare. Gothenburg's trams cross Kungsgatan at two points, Stockholm has a major subway station directly under Sergelgatan, while most of Orebro's bus lines serve Drottninggatan, the transit mall. The transit stops generate additional pedestrian traffic on the malls, as is evidenced by Linkoping, where 26,000 pedestrians/day were counted where Nygatan passes by the main bus stop; by comparison, the next most active location had 17,000 pedestrians/day (Linkoping Planning Department 1989).

Conversely, transit vehicles, particularly buses, can exert a negative impact on pedestrians. The tremendous volume of buses that interface with the mall (i.e., 1100 buses/day cross Nygatan in Linkoping, and 1400/day use Orebro's transit mall) result in significantly high levels of air and noise pollution. Furthermore, one has to be very careful when crossing these busways, thereby detracting from the quality of the pedestrian experience. This problem is so serious that planners in Orebro, Linkoping, and Uppsala have recommended that the number of buses interfacing with pedestrian streets be greatly reduced.

Despite the large number of Swedes who commute by bicycle and transit, it is difficult to ignore the rapidly increasing popularity of the automobile. Sweden has the third highest rate of automobile ownership in Europe, trailing only West Germany and Switzerland (Pucher 1990). A paradox exists for many environmentally-conscious Swedes: A growing desire to use automobiles versus the increasing awareness of its inherent problems. The continual increase of automobiles penetrating city centres diminishes the safety and environmental qualities of the pedestrian experience, thus contradicting the traffic management goals discussed above. Recognizing the severity of this problem, Uppsala, Gothenburg, and Stockholm are studying ways in which to further restrict cars from their centres (i.e., car tolls, increased parking costs, more pedestrian-only streets); in fact, the city council of Gothenburg is deliberating over a plan to make the entire central core virtually free of automobiles.

Public and private officials interviewed in the six cities did agree on one point: Adequate and convenient parking is the key for pedestrian streets remaining viable. While it is not important for customers to be able to drive in front of a store, as merchants once believed, it is essential that customers be able to park within reasonable walking distance (general consensus was within 200 meters). Thus, plans to erect additional parking ramps within a block or two of the pedestrian malls are being made in each city examined.

Design features

In general, Swedish pedestrian streets do not have an abundance of urban design amenities. The design schemes can best be described as basic, hard-edged, and simplistic. There exists a minimal amount of landscaping (i.e., trees, flowers, shrubs, grassy areas) or other design amenities, such as fountains and public art. Even the pavement covering the street tends to be utilitarian and lacking in aesthetic interest (see Figure 4.3).

The lack of seating is the most conspicuously absent design feature. In most of the cities visited, there was little seating provided directly on the pedestrian street. In Uppsala, Orebro, and Falun, the only places to sit were the shaded and crowded bus shelters, while Stockholm and Gothenburg did not even provide these seating possibilities for walkers. Gothenburg does have many outdoor cafes and restaurants, but one has to be a paying customer to sit in one of these establishments. Having a place to sit and relax is desirable not only to children and the elderly, but to the average pedestrian who is tiring from shopping, waiting to meet someone, or simply wanting to look at other people.

The most apparent reason for this minimalist design approach is that the streets are too narrow and crowded with people to contain amenities such as benches, landscaping, and fountains. This is especially true for pedestrian streets in the larger cities of Stockholm and Gothenburg, where certain segments of Drottninggatan and Kungsgatan, respectively, already handle too much pedestrian traffic for additional amenities to be included. In smaller cities, however, due to the lower densities of people, the potential for improved design features is much greater. In fact, the mid-size cities of Vastaras and Kalmar reportedly have very attractively landscaped malls.

Another school of thought contends that additional design amenities are unnecessary on Swedish pedestrian streets because the people themselves constitute the most essential design feature (Nordquist 1984). The primary element that attracts people to use an urban space, according to this philosophy, is the presence of other people, a feature not in short supply on the walking streets visited. Still, others believe that the inclusion of seating and landscaping would add significantly to the quality of the pedestrian experience, although perhaps not to the actual quantity of people.

Linkoping's system of pedestrian streets does not conform to the general design scheme described above. The well-integrated system of streets and squares is attractively landscaped with trees and plants, and provides numerous seating opportunities, especially where the streets enter the squares. Lillatorget is arguably the best designed space studied, as it contains a sculptured rock fountain, large trees providing welcome shade, ample seating, a large and popular outdoor cafe, a two-tiered walking surface, and a covered walkway and additional benches on Replagaregatan approaching the square. Not surprisingly, this is a popular and widely-used city space (see Figure 4.4). All three squares serve as collection points for the streets designed to move people, and tend to be where people congregate and spend time.

The architecture of buildings that front the pedestrian streets is oftentimes uninspiring. With the exception of Vasterlanggatan in Stockholm's Old Town and Tanneforsgatan in Linkoping, there are very few pre-20th Century structures lining these streets. Instead, most of the systems primarily possess uninteresting retail and office buildings constructed since the second world war. Several of the subjects interviewed commented on how obvious it was that post-1960s buildings have been designed with little thought given to how well they integrate with the pedestrian street.

Several other design features cause concern. First, each pedestrian system was located within one block of a body of water, yet only Stockholm provided a direct pedestrian link; given the way that water attracts people, a tremendous design opportunity has been overlooked. Second, a pair of second-level double-decker skywalk bridges cross Fredsgatan in Gothenburg. These bridges create dark, gloomy spaces underneath, and tend to remove people from the street-level. Third, several large dead spaces were observed, usually where a non-commercial structure interrupts the shopping promenade, such as a library in Uppsala and a corporate headquarters in Falun. In general though, dead spaces were rare on Sweden's walking streets. Finally, an increasing number of indoor gallerias have been constructed recently in city centres throughout Sweden. Since shop entrances are located off interior courts, not the street itself, gallerias can work to decrease the number of people on the pedestrian streets (Gehl 1989).

Retail activity

The majority of people interviewed thought that pedestrian streets have helped to strengthen the commercial cores of Swedish cities. Today, most downtown merchants embrace the idea of expanding central pedestrian districts, and in some instances, have even initiated the idea themselves. This support from the shopkeepers is somewhat ironic because they strongly opposed the concept of pedestrian streets when initially proposed. For example, "Save the Main Street" was the merchant's battle cry when Uppsala was first contemplating this strategy in the late 1950s. Obviously, this shift of opinion is due to the recognition that pedestrian streets are good for commerce.

The inventory of the six pedestrian systems revealed that 77.1% of street-level establishments were engaged in the sale of retail goods, most commonly clothes, shoes, and jewelry. The domination of retail stores ranged from 67.1% in Falun to 82.6% in Stockholm (see Table 4.2). Of note, based on the case studies, it appears as though larger cities have a higher percentage of retail shops on their pedestrian streets than do smaller cities, probably due to the fact that only retail establishments can afford the expensive mall space in larger cities.

Large department stores have dominated Swedish pedestrian malls since their inception. Stores such as H & M, Domus, and Ahlens were found on most of the systems, usually in strategic locations. During the 1980s, however, department stores have lost some of their luster, and are gradually losing their customers to chain specialty stores. The customer base for downtown department stores has remained stable at best and, in some instances, has declined; for example, Domus has been downsized in both Falun and Linkoping.

Besides the changing status of department stores, the retail mix on pedestrian streets has undergone other changes since their inception in the 1960s. The types of shops rarely seen today are those selling larger items which usually require an automobile to transport home (i.e., groceries, furniture, appliances, hardware). The disappearance of grocery stores from downtown has been especially noteworthy. Due to the convenience provided by suburban supermarkets that cater to the automobile, several department stores have recently discontinued carrying groceries, such as Domus in Linkoping. Evidently, pedestrians are unable to purchase sufficient amounts of groceries to be profitable. Not only have shops selling certain types of goods left, but chain stores have been replacing local independently-owned shops in city centres throughout Sweden. Consequently, the retail mix looks very similar from city to city.

The other three categories of establishments presented in Table 4.2 are far less represented than are retail goods. Places serving food/drinks (i.e., restaurants, cafes, pubs, fast food stands) constitute less than 10% of the establishments in every city except Gothenburg. Establishments providing personal services (i.e., travel agencies, hair salons) and business services (i.e., banks, real estate, investment brokers) generally represent less than 6% of the pedestrian street facilities. Shops and services from these three categories, with the exception of banks, frequently were found on streets parallel or adjacent to the pedestrian street, where the rents usually were lower.

The presence of vendors, a sometimes controversial element on pedestrian streets, varied considerably from city to city. No vendors were observed in Orebro or Gothenburg, while a few licensed stalls and carts were found in Falun and Uppsala. Several days each week, two of Linkoping's squares are converted into lively marketplaces full of stands selling everything from fruit to flowers to fabric. In Stockholm, a host of unlicensed vendors selling a wide variety of jewelry and accessories situated themselves on Drottninggatan's busiest intersections; the police occasionally would force them to vacate, but the resourceful vendors would return shortly. Shopowners habitually complain about vendors, but they do tend to add vitality to the street.

My research revealed three other retail-related problems that need to be addressed. First, the activity generated by pedestrian streets often is at the expense of shops located on other downtown streets, leaving the impression at times that these non-pedestrianized streets are merely "back alleys" (Nordquist 1984). The popularity of one pedestrian street can even hurt retail sales on another pedestrian street; in Falun, shops along Asgatan are gradually dying due to the concentration of activity on Hotmgatan.

Second, the rapid growth of chain stores in Sweden's city centres is forcing many small independent shopowners, especially those providing basic services, to close or relocate. Often the relocation is to less desirable streets on the periphery of the centre. This trend impedes the ability of pedestrian streets to provide a variety of shops and services. Berdichevsky (1984) and Hajdu (1988) are critical of this trend in Denmark and West Germany, respectively. Berdichevsky (p. 20) states that pedestrian streets "...have become 'consumer corridors' specializing in the luxury goods of an affluent consumer society." The domination of retail stores at the expense of other types of establishments, as seen in Table 4.2, clearly indicates the prevalence of this trend in Sweden.

This lack of variety manifests itself in a third problem: The absence of nightlife on all six pedestrian streets. Once the shops close around

6:00 p.m., not many establishments remain open for business, save a few restaurants and the occasional movie theatre (of note, Orebro does have a 10-plex cinema on Drottninggatan). Not one establishment offering musical entertainment was observed on any of the pedestrian streets. Hence, the number of people thins out quickly as evening approaches because there is little to attract them.

Key issues for the 1990s

A number of key issues facing Swedish pedestrian streets in the 1990s were identified through my research. First, a better balance and variety of establishments needs to be achieved along many pedestrian malls. The replacement of some retail shops selling similar goods with restaurants, cafes, and establishments offering personal services would help to diversify the pedestrian street and create more reasons for people to go there. The increasing domination of chain stores leads to a sameness of malls from one city to the next; thus, cities should try to encourage more independent, locally-owned establishments to locate on the mall. The replacement of some shops with cafes and nightclubs would help promote evening activity as well, thereby extending the hours pedestrian streets remain vital.

A second key issue is accessibility. If pedestrian streets are not easily accessible to a large segment of the local population, their activity levels will steadily diminish. In larger cities, given the large numbers of people coming to the city centre, the answer is to integrate fast and efficient transit service with the pedestrian system. In Stockholm, a major subway station situated under Sergelgatan serves as an excellent transit-pedestrian linkage, while Gothenburg is planning to modernize its fleet of trams so as to serve the pedestrianized centre more effectively. In mid-sized and smaller cities, accommodating the automobile driver is the key to accessibility. This means that additional parking ramps need to be built in close proximity to the pedestrian system; the parking also must be evenly distributed throughout the system, or those segments not near to parking will suffer disproportionatley.

Third, a general consensus of those interviewed suggests that more attention needs to be given to design details. As is the case throughout much of

Europe, the design of most Swedish malls can be described as noninnovative. This may have been sufficient in the 1960s when conservation and traffic management were the primary reasons for converting these streets. But today, with stiff retail competition coming from suburban shopping centres, it is crucial for city centres to offer consumers an enticing, exciting, and attractive pedestrian environment, one that is difficult to duplicate in the suburbs. Therefore, by providing more quality landscaping, seating, lighting, fountains, and public art, and by linking malls with nearby natural waterways, the pedestrian system will become more beautiful and user-friendly. The fact that all of the proposals for redesigning Drottninggatan in Orebro call for a wide range of landscaping and design amenities suggests that design features on malls will become more prominent during the 1990s.

A fourth trend which concerns many public officials is the proliferation of indoor gallerias in city centres. While these developments do add to the total retail space, they also possess the potential to draw shops and customers away from the pedestrian street itself. Mini-gallerias, with perhaps 10-15 establishments, were found in cities large and small throughout the country. The fact that these spaces are enclosed and privately-controlled suggests that they contradict the spirit and essence of public streetlife which the pedestrian street provides. In the case of Nordstan, a huge galleria containing over 130 stores in Gothenburg's centre, many of the current establishments moved to the galleria from Kungsgatan. If future gallerias are to be erected, the key may be to design them so as to integrate with the street, rather than compete with it.

An overall concern which may encourage city officials and the downtown business community to pay special attention to the above issues is the intense completion provided by suburban shopping facilities which are easily accessible by automobile. The key to dealing with this competition may not be simply to accommodate the automobile, although improved parking facilities are essential, but rather to make the overall environment more oriented to the needs and comfort of the pedestrian. This notion may actually require further restrictions on the automobile. To achieve this, planners should try to design the overall downtown transportation system so as to minimize the instances where people and vehicles interface. For example, when walking on the pedestrian streets of Linkoping and Uppsala, one is barely cognizant of motor vehicles because cars do not intersect walking zones. Conversely, in Stockholm, and to a lesser extent Gothenburg and Orebro, one has to be constantly aware of crossing traffic, thereby detracting from the quality of the pedestrian experience.

Concluding remarks

The above issues need to be addressed if pedestrian streets are to remain economically viable, thus implying that commercial development will emerge as one of the key goals for malls during the 1990s. Still, one cannot deny the continuing popularity that these streets have achieved in city centres throughout Sweden. While a myriad of factors help to explain this popularity, one reason stands out: The role of the central meeting place, traditionally important in Swedish culture, has been assumed by the pedestrian street system in most of Sweden's cities. The pedestrian street will continue to serve as the place where people meet and where they go to see and be seen. Nevertheless, the growing use of the automobile dictates that for pedestrian streets to maintain this role, a balance somehow must be struck between the demands of the automobile driver and the social, cultural, and physical requirements of pedestrians.

Note: Many of the Swedish words used in this chapter, including the cities of Orebro and Linkoping, contain Swedish characters not available to the typesetter. In their place, the author has substituted the closest English vowel, although it is recognized that this is an inaccurate spelling of the words.

Table 4.1: Background Information on Swedish Case Studies

City	1988 City Population	Major Pedestrian Streets	Year Opened[a]
Falun	52,448	Staaggatan Asgatan Hotmgatan	1969
Gothenburg (Goteborg)	430,763	Kungsgatan Fredsgatan Korsgatan	1961
Linkoping	119,167	Tannesforsgatan Nygatan Repslagaregatan	1967
Orebro	119,824	Kopmangatan Drottninggatan (transit)	1972
Stockholm	669,485	Vasterlanggatan Drottninggatan Sergelgatan	1961
Uppsala	161,828	Kungsangsgatan Svartbacksgatan	1964

[a] Sources: Gunnarsson and Lindstrom (1969) and interviews

Table 4.2: Inventory of Establishments Fronting Pedestrian Streets - 1990

City	Total Establishments	Retail Goods	Business Services	Personal Services	Food/ Drink	Other
Falun	76	51 (67.1%)	10 (13.2%)	7 (9.2%)	5 (6.6%)	3 (3.9%)
Gothenburg	127	99 (77.9%)	2 (1.6%)	5 (3.9%)	20 (15.7%)	1 (0.8%)
Linkoping	107	79 (73.8%)	7 (6.5%)	9 (8.4%)	10 (9.3%)	2 (1.9%)
Orebro	108	78 (72.2%)	10 (9.3%)	9 (8.3%)	6 (5.6%)	5 (4.6%)
Stockholm	242	202 (82.6%)	8 (3.3%)	8 (3.3%)	19 (7.8%)	5 (2.1%)
Uppsala	66	51 (77.3%)	6 (9.1%)	2 (3.0%)	5 (7.9%)	3 (4.5%)
TOTALS	726	560 (77.1%)	43 (5.9%)	40 (5.5%)	65 (8.9%)	19 (2.6%)

Notes: 1. Inventory conducted during spring of 1990
2. Totals do not include vacancies or buildings under construction

Figure 4.1: Uppsala Mall. Despite increased competition from suburban shopping centers, activity levels on pedestrian streets such as Kungsangsgatan remains fairly high (1990)

Figure 4.2: Stockholm Sergelgatan. Sergelgatan was a brand new pedestrian street created as part of an urban renewal plan in the 1960s (1990)

Figure 4.3: Orebro Mall. Typical of most Swedish malls, there is no landscaping of design amenities on Kopmangatan (1990)

Figure 4.4: Linkoping Square. The well-designed Lillatorget provides a variety of seating options (1990)

5 Downtown pedestrian malls in Sweden and the United States: A comparative study

During the 1950s and 1960s, city centres on both sides of the Atlantic were experiencing difficulties. A fundamental cause of these problems was the increasing use of the automobile, which resulted in the decentralization of activities from North American downtowns and in traffic and environmental concerns for European city centres. A commonly-applied strategy used to address this problem was the pedestrian mall. Despite similar origins, North American pedestrian malls today exhibit notable differences when compared to their European counterparts, and are perceived to be less viable by most observers.

Sweden and the United States constitute two nations in which pedestrians streets have been widely implemented. While the malls in each country share numerous similarities, they also demonstrate significant differences, especially in terms of their activity levels. The status of pedestrian streets today perhaps is the most important variation between the two nations. Most Swedish malls are perceived to be successful, although changes in transportation and shopping preferences are causing some problems. In America, however, many pedestrian malls are considered failures to the point where an increasing number are being converted back to thoroughfares allowing automobiles (Houstoun 1990).

Utilizing Sweden and the United States as representative nations, the purpose of this chapter is describe, compare, and evaluate pedestrian malls in Europe and North America on the basis of their historical development, design features, retail trends, and transportation considerations. The comparisons are based on the findings reported in chapters 3 and 4. Through a discussion of the key differences and their causal factors, planners on both continents should be able to gain valuable insights regarding the most significant features contributing to mall success and failure, and thus better understand the forces that are shaping pedestrian malls with which they are familiar.

Pedestrian mall development

The timing of pedestrian mall development was very similar in both countries. Kalamazoo (Michigan) implemented the first American downtown mall in 1959. Over the following fifteen years, nearly 200 American cities followed Kalamazoo's lead, especially small and mid-sized cities. Mall development peaked in the mid-1970s, however, and only a handful of new pedestrian streets have been constructed during the 1980s. In Sweden, where the influence of early Dutch and German malls did not go unnoticed, the concept of pedestrian streets had been contemplated as early as 1955, when it was included as part of Uppsala's master plan. Due to the strong resistance by downtown shopkeepers, this idea was not implemented in Uppsala until 1964. However, several other Swedish cities, such as Lund, Stockholm, Gothenburg, and Helsingborg, did apply this strategy in 1961, and by the end of the decade, 35 cities possessed pedestrian streets in their city centres. Today, experts estimate that over 100 Swedish cities have downtown malls, most of which exhibit high activity levels.

While the development of malls was occurring simultaneously, the reasons for their implementation differed between the two countries. The primary reason in most American cities centered on the economic revitalization of downtown. Downtowns, heretofore the unchallenged leader of metropolitan retailing, were losing customers and businesses to suburban shopping centres, which offered consumers free parking, a convenient location, and a safe and pleasant shopping environment. Pedestrian malls became the symbol of hope to many city planners and downtown officials because of their potential to provide a more aesthetic and human-scale downtown that would be attractive to shoppers.

The need for improved transportation management was the major factor underscoring mall development in Sweden. The increasing use of private automobiles was clearly incompatible with the economic, cultural, and physical fabric of city centres, which were oriented towards the needs of pedestrians and public transit. Consequently, the narrow streets were clogged with motor vehicles, noise and air pollution approached critical levels, walking was becoming increasingly dangerous, and a "way of life" characteristic to city centres for generations was in great peril. Therefore, in many Swedish cities, comprehensive traffic management plans were developed that recommended pedestrian-only streets as a way to promote the separation of traffic in city centres.

The configuration of malls also exhibited some notable differences. The typical American mall consists of a few linear blocks along the traditional downtown "main street". There are three basic types of downtown pedestrian malls in America: the traditional pedestrian street, the shared mall, and the transit mall (see chapter 2 for more detailed definitions). Mall configuration in Sweden is more difficult to categorize. As in America, most malls are linear in orientation and are located along the traditional shopping street. Many Swedish city centres, however, contain a series of interconnected pedestrianized streets that together constitute a central pedestrian street system (see Figure 5.1). Linkoping, for example, possesses a well-integrated system of squares and walking streets that actually extends beyond the commercial core into nearby residential areas. Transit malls are rare in Sweden--Orebro being a notable exception--and no examples of shared malls were uncovered.

Comparing design characteristics

Greater attention is paid to design amenities and landscaping on American pedestrian streets than for their Swedish counterparts. For most malls in the United States, an attempt has been made to provide attractive landscaping through the use of trees, shrubs, flowers, and grassy areas (see Figure 5.3). Moreover, the use of other design amenities, such as benches, fountains, public art, and handsome pavement, is commonplace. In fact, some observers have even criticized American malls for being over-designed, with their dense foliage and oftentimes cluttered look.

Conversely, Swedish pedestrian streets generally do not possess an abundance of urban design amenities, a characteristic common to malls throughout Europe (Brambilla and Longo 1977). Rather, their design schemes can best be described as basic, hard-edged, and simplistic. Most of the pedestrian streets contain no landscaping or street furniture (see Figure 5.4). The lack of seating is the most troublesome and conspicuously absent design feature. Besides the shaded and usually crowded bus shelters available on some malls (i.e., Orebro, Uppsala, Falun), few public benches were observed directly on pedestrian streets. Having a place to sit and relax is desirable not only to children and the elderly, but to the typical pedestrian who is tiring from shopping, waiting to meet someone, or simply wanting to people-watch.

A design-related objective of both American and Swedish pedestrian streets is to increase the quantity and quality of downtown open space. The mall itself contributes to the amount of open space available. American malls generally have a softer, more park-like appearance due to the emphasis placed on landscaping and design. However, since the volume of space provided by these linear corridors is usually insufficient, the key to satisfying this objective is to provide ample and well-integrated public spaces adjacent to the mall corridor. In this regard, Swedish malls have a clear advantage because most are linked with a series of adjoining squares (most notably, Linkoping, Orebro, and Falun); most American downtowns do not possess public squares due to their grid orientation. The only American city visited that addressed this problem effectively was Minneapolis, where several well-designed public plazas have been created next to Nicollet Mall.

A design problem that is particularly noticeable in the United States is the amount of dead space found on malls. These dead spaces serve to interrupt the wall of continuous activity that originally characterized commercial pedestrian streets; this continuous wall not only consisted of shops and restaurants, but could also include outdoor cafes, vendors, open spaces, or any other activity that generated human interaction. Vacant storefronts and surface parking lots are two examples of dead spaces that create negative visual impressions and act to impede pedestrian flow and spirit. Some dead spaces result from insensitive building design, where blank walls fail to relate to street-level. A classic example is in downtown St. Cloud, where a suburban-style structure was erected on Mall Germain in 1985; this building, constructed to house a major department store, has no street-level windows or shops facing the mall. Dead spaces were far less frequent on Sweden's pedestrian streets.

Comparing commercial trends

The inventory of establishments fronting pedestrian malls reveals that stores selling retail goods dominate commercial structure in both countries; however, this dominance is even more extensive on Swedish malls. As reported in Table 3.2 and 4.2, 61.1% and 77.1% of establishments with entrances directly on the malls (shops in indoor gallerias were not counted) were engaged in the sale of retail goods in the United States and Sweden, respectively. Shops selling shoes, jewelry, and women's apparel were the most common. Malls in Sweden consistently demonstrated a higher share of retail shops, as five of the six Swedish cities displayed a greater percentage than did the highest American city (Denver - 69%).

Clearly, pedestrian malls are shopping streets in Sweden, whereas American malls exhibit a bit more variety in the types of services rendered. A higher percentage of American establishments are engaged in the three categories other than retail goods presented in Tables 3.2 and 4.2 than is the case in Sweden. While the relative number of places offering personal services (i.e., travel agents, hair stylists) and business services (i.e., financial institutions, real estate agents) was small in both countries, a significant difference exists in the number of establishments serving food/drinks, such as restaurants, cafes, pubs, and fast food outlets. The percentage was over twice as high in the United States (20.2%) than in Sweden (8.9%). On every American mall, at least 13% of the businesses served food/drinks, while less than 10% of the mall establishments are involved in this activity in each Swedish city except Gothenburg. In both nations, establishments from these three categories, with the exception of banks, frequently are found on streets parallel or adjacent to the mall, where rents usually are lower.

The centerpieces for early American and Swedish malls were the large department stores that usually occupied a prime location. Over the years, their role and status has changed. In Sweden, they have lost some of their luster, and are gradually losing customers to chain specialty stores. Downtown department stores in Linkoping and Falun, for example, have been downsized in recent years, and virtually no expansion is planned in city centres for the near future. Nevertheless, stores such as Ahlens, H & M, and Domus remain as key fixtures on Swedish pedestrian malls. The status of department stores in American downtowns is more mixed. Providence, which has lost three of its four department stores during the past twenty years, is representative of many cities where the declining downtown has been unable to support these large emporiums. On the other hand, Minneapolis, despite losing two stores in the 1980s, has been successful in attracting two new prestigious department stores--Saks 5th Avenue and Neiman Marcus--to Nicollet Mall.

Three other features of commercial structure are shared by the two nations. First, there was a sudden dropoff of mall activity levels after 6:00 p.m. because few shops remain open in the evening and because of the virtual nonexistence of places offering nighttime entertainment (i.e., bars, nightclubs, theatres). Madison (Wisconsin) is a notable exception to this trend. Second, a recent trend has been the construction of indoor shopping gallerias where shops are entered from interior courts, not from the street itself. These indoor shopping centres are very popular in city centres of both the United States and Sweden, and have the potential to keep people off of the mall, thereby hurting street activity levels and the economic viability of mall establishments. Third, stores that are affiliated with national chains are gradually replacing the independent locally-owned enterprises that once dominated malls. Many of the establishments being replaced offered basic goods and services, thus their loss has hurt retail diversity.

Retail sales on central city malls in both

countries are being threatened by the increasingly popular suburban shopping centre. Thus, the ability for pedestrian streets to help downtown retailers compete with suburban stores is extremely important today. Most of the people interviewed in Sweden firmly believed that malls have helped to strengthen downtown commerce; even the shopkeepers, who in the 1960s strongly resisted the notion, now embrace the idea of expanding pedestrian districts. This attitude is far less prevalent in America, where public officials and the downtown business community are more enamored with pedestrian skywalks, large indoor shopping gallerias, and convention centers as downtown redevelopment strategies than with pedestrian malls. Many are convinced that existing malls should converted back to streets allowing automobiles, as has happened in Providence and other cities.

Comparing the transportation dimension

Accessibility is a key factor in determining the success and viability of pedestrian malls. According to interviews conducted in both countries, malls need to be easily accessible to a large segment of the local and regional population in order to maintain stable activity levels. Oftentimes this accessibility is a function of well-integrated linkages between the mall and various modes of transportation used in the city centre, such as bicycles, transit, the automobile, and other pedestrian systems. Overall, the integration of other transportation modes with the pedestrian streets is far more developed in Sweden than in America.

Nearly one-half of urban passenger trips in Sweden, according to Pucher (1990) are made through non-motorized means--pedestrian and bicycle--compared to only one in ten American trips (see Table 5.1). Consequently, it should not be surprising that Swedish malls are better integrated with pedestrian and bicycle transportation. Linkoping and Gothenburg provide examples of Swedish cities where the commercial pedestrian streets are only one part of a larger pedestrian district that encompasses much of the city centre. Furthermore, the malls are often designed so as to minimize the frequency wherein pedestrians and motor vehicles compete for space (e.g., Uppsala, Linkoping).

Conversely, the only street-level pedestrianized area in most American downtowns is limited to the mall itself. The proliferation of second-level skywalk systems in many American cities, including Minneapolis where over 40 downtown blocks are connected, has resulted in a competing pedestrian system which usually detracts from mall activity levels.

Bicycles are used for both commuting and recreational purposes by a majority of Swedes. For example, city officials estimate that over 15,000 people enter Orebro's city centre by bicycle on a typical summer day; many of them ride on Drottninggatan, which doubles as a transit mall and bicycle pathway. To accommodate bicyclists, convenient and ample bicycle parking facilities are provided on and/or adjacent to the mall in Falun, Linkoping, Orebro, and Uppsala. These racks were heavily used throughout the day. Given the limited popularity of bicycles as a commuting vehicle in the United States, little attention is given to the needs of bicyclists on American malls. The one exception is Madison (Wisconsin) where, due to the large number of college students using the downtown, many bicyclists ride and park on State Street Mall.

The linkage between pedestrian streets and mass transit was clearly evident in each of the six Swedish cities. The major downtown bus terminals in Uppsala, Falun, and Linkoping are located on central squares where the mall is crossed by a bus-only street. Gothenburg's trams intersect the mall at two points, Stockholm placed a major subway station directly under one of its key pedestrian streets, while most of Orebro's bus lines utilize the transit mall. These transit transfer points generate additional pedestrian traffic for the malls, as is evidenced by Linkoping, where 26,000 pedestrians/day were counted where the mall passes by the main bus stop; by comparison, the next most active location had 17,000 pedestrians/day (Linkoping Planning Department 1989).

The linkage pattern is less consistent on American malls. Buses do not interface directly with the mall in Providence, Santa Cruz, or St. Cloud, thus forcing passengers to walk several blocks to make their connections. On the other hand, Minneapolis, Denver, and Madison possess transit malls that bestow greater mall accessibility to bus passengers than to automobile drivers. Denver's innovative 16th Street Mall, with its team of electrically-powered shuttlebuses connecting transit terminals located at each end, is especially noteworthy (see Figure 3.5). It is interesting that transit malls are much more common in the United States despite the fact that transit is more widely-utilized in Sweden. A number of successful American malls possess a transit element to help maintain consistently high levels of mall activity that evidently cannot be achieved through pedestrians alone; this need to enhance activity levels is unnecessary in Sweden.

Public officials in both nations acknowledge the tremendous importance of the automobile in downtown commuting patterns. Pucher's (1990) study of automobile ownership not only demonstrates the well-known fact that Americans have the highest rate in the world, but that Sweden has the third highest rate in Europe. Thus, the provision of adequate and convenient parking facilities close to the mall, always a key to mall success in American downtowns, is increasingly becoming important to the future viability of Swedish pedestrian streets as well. In the United States, the desire to cater to the needs of automobile drivers has reached the point where many cities, including Providence, have reconverted their malls back to thoroughfares allowing vehicular traffic. To date, there were no reports of mall reconversion in Sweden.

Explaining Swedish-American differences

To summarize, American pedestrian malls exhibit a number of significant differences when compared to their counterparts in Sweden, including a) the original goals centering more on economic redevelopment than traffic management, b) lower activity levels and general public acceptance, c) a more linear configuration focusing on a single street, d) more skepticism by the downtown business community regarding mall effectiveness, e) a greater emphasis on design features and landscaping, f) more dead spaces, g) a more varied retail mix, especially in terms of establishments serving food/drinks, h) a lesser degree of integration with other modes of transportation, and i) a more widespread application of transit malls. A combination of factors help to explain these differences between pedestrian streets in the United States and Sweden specifically, and in North America and Europe generally.

First, the timing of commercial suburbanization came much later in Sweden than in America. Competition from suburban shopping centres was a fact of life for most American downtowns by the early 1960s, thereby creating the need for strategies such as pedestrian malls that were designed to help attract consumers downtown. Because suburban competition was not a major concern in Sweden until the 1980s, issues related to convenience (i.e., parking, store hours) and aesthetics (i.e., design amenities) have only recently become significant. As more Swedes own and use automobiles, it will become increasingly important for pedestrian streets to be accessible to drivers and to offer an interesting and aesthetic experience not provided in the suburbs. The design of American malls reflects the fact that downtown planners have long been cognizant of their suburban competition. The dominance of stores selling retail goods on Swedish pedestrian streets is another result of this subsequent suburbanization; given the demand for downtown space, retail shops were more willing and able to afford the premium rents being charged than were other enterprises.

Second, the form and structure of Swedish cities varies considerably from their American counterparts. Most city centres in Sweden are oriented around public squares and plazas that provide natural open space and a place into which malls can funnel. Conversely, American cities tend to be laid out in rectangular grids without squares. Therefore, it is more necessary that American malls be landscaped so as to create parklike corridors. Also, the American streets are much wider, thereby allowing more opportunity for providing design amenities; many Swedish malls are far too narrow to permit space-consuming landscaping and street furniture. The greater density and compactness characteristic of Swedish downtowns helps to promote the high activity levels attributed to pedestrian streets. Many American downtowns are simply too spacious to facilitate an active and festive streetlife. Finally, as discussed in Martin (1990), Europeans have less private space than do Americans, and therefore are more likely to utilize public spaces such as malls for social interaction.

Third, a more comprehensive approach to urban planning characterizes the development of Swedish cities. Most Swedish cities are guided by a coordinated traffic management plan which attempts to integrate the needs of all forms of transportation, including pedestrianism. Consequently, pedestrian malls are planned in conjunction with bicycles, transit, highways, and parking. In many American cities, separate plans are developed for each mode of transportation, and the linkages between them are too often an afterthought. Furthermore, Swedish malls appear to be better integrated with the downtowns in which they are located in terms of housing, open space, recreation, and transit, again pointing to a more comprehensive downtown planning process than is the case in the United States.

Fourth, social and cultural factors may best explain why pedestrian streets are more popular in Sweden than in the United States. The role of the central meeting place, traditionally important in Swedish culture, has been assumed by pedestrian malls in Sweden. It serves as the place where people meet, interact, and go to see and be seen. Americans do not possess this tradition, instead preferring to interact in more private settings. The cultural homogeneity of the Swedish people, when compared to the more pluralistic Americans, also may contribute to their higher level of comfort in crowded public spaces (Martin 1990). Finally, the act of walking, whether for commuting, social, or recreational purposes, is far less developed in American society than in Sweden or any other European nation. Is it any wonder that pedestrian malls in the United States have fallen upon hard times when many Americans find it difficult to understand why Europeans actually choose to stroll down city streets even when the shops are closed?

Concluding remarks

This chapter has emphasized the key differences between Swedish and American pedestrian streets. Yet, as we enter the final decade of the 20th Century, mall planners in both nations will need to address some common issues. First, how can pedestrian malls specifically, and downtowns in general, remain competitive with suburban retail centres? This problem has recently become a major concern to Swedish downtowns, just as it has been to American downtowns for decades. Second, how can city centres better accommodate the automobile without sacrificing the pedestrian and transit advantages historically inherent to commercial cores? For instance, if improved parking facilities are provided, thereby encouraging more cars to enter downtown, will the quantity and quality of pedestrian spaces be reduced? Third, how can the projected increase of indoor gallerias be integrated into downtowns without draining the street of pedestrian vitality? The design of most downtown gallerias results in these complexes being competitive, not complimentary, to pedestrian malls. Finally, and perhaps most importantly, what can planners do to enhance the overall quality of the downtown walking experience so as to attract more people to choose to be in the city centre?

These issues are presented as questions because, as most planners involved to downtown development know, there are no easy or definitive answers to be provided. It is for this reason that planners are encouraged to look beyond their own borders so as to understand the dynamics of malls in other countries. No one is suggesting that pedestrian streets in one country should try to emulate those in another; as presented above, the prevailing social, spatial, and cultural differences, especially when comparing American and European societies, make this idea absurd. However, valuable lessons can be learned through the realization that there are other approaches to mall planning and design, some of which may be applicable at home.

Note: For comment on spelling of Swedish words, see end of Chapter 4.

Figure 5.1: Falun Mall. Hotmgatan connects one square (right) with a second square (far center), thus providing an example of a well-integrated pedestrian system consisting of a series of streets and squares (1990)

Figure 5.2: Denver movable chairs. The innovative and attractively landscaped 16th Street Mall today serves as the prototype transit mall in America (1989)

Figure 5.3: Providence Westminster mall

These two photographs illustrate the fundamental design differences between American and Swedish malls. Westminster Mall in Providence is nicely landscaped and adorned with a variety of street furniture, while the much narrower Nygatan in Linkoping contains few of these design amenities.

Figure 5.4: Linkoping mall

6 Pedestrian skywalks in American downtowns

Elevated pedestrian skywalks are increasingly becoming a popular element of the downtown streetscape. During the past two decades these systems have been constructed in numerous downtowns, particularly cold weather cities of Canada and the American Upper Midwest. The most extensive networks have been constructed in medium-sized northern cities such as Calgary, Des Moines, Edmonton, Milwaukee, Minneapolis, Spokane, St. Paul, and Winnipeg. Smaller cold weather cities in the midwest, including Cedar Rapids (Iowa), Duluth, Fargo, Lincoln, and Sioux City (Iowa), have built systems as well, as have an increasing number of cities that enjoy milder climates (i.e., Charlotte, Dallas, Ft. Worth, Cincinnati, Portland OR). The underrepresentation of very large cities or cities located on either coast is noteworthy.

Skywalks may represent harbingers of the 21st Century downtown, which is likely to become increasingly indoor-oriented and less dependent on the street-level for its vitality. Besides the physical manifestations, skywalks are likely to contribute to a major economic and social transformation of downtown space. Economically, they can cause significant shifts in where and how people shop, entertain, and conduct business, thereby affecting land use and investment patterns. Socially, skywalks constitute yet another means through which people can isolate themselves from social environments that differ from their own.

Therefore, with skywalk systems playing a leading role, not only may the streetscape evolve into a pedestrianless urban space, but the very essence of the downtown itself may be dramatically altered.

A skywalk system can be defined simply as a network of elevated interconnecting pedestrian walkways. The network consists of bridges over streets, second-story corridors within buildings, and various activity hubs (i.e., indoor shopping gallerias, hotel or office building atriums). For the most part they are enclosed and climate-controlled, and connect a body of retail and service establishments that may or may not have any relationship to the street level.

The notion of connecting buildings above street-level is not exactly new. Elevated bridges that connected parts of castles, palaces, and other major buildings dating back centuries can be found throughout Europe; the same can be said for certain American factories and office buildings constructed during the 19th and 20th Centuries. Even the concept of a system of elevated walkways predates the current skywalk configurations. Futuristic urban schemes, ranging from Leonardo da Vinci's 15th Century city designs to Moses King's 1908 Views of New York to Le Corbusier multi-layered city, have all incorporated upper-level passageways to protect pedestrians from the weather, animals, and street vehicles (Rowe 1985). Cincinnati's 1925 City Plan considered the inclusion of "overhead sidewalks", but decided against their implementation because of the negative impact they might exert on the street-level (Forusz 1981). These historical precedents notwithstanding, it is generally acknowledged that the first successfully implemented skywalk system opened in Minneapolis in 1962 (see Figure 6.1).

Several reasons help explain why many cities have chosen to employ skywalk systems in their downtowns. First, they facilitate pedestrian mobility, allowing most trips to be made in less time and in more comfort, especially in bad weather. Second, since pedestrians are separated from vehicular traffic, improved pedestrian safety is accommodated. And finally, skywalks are being used as a downtown redevelopment tool because of their ability to promote density, create a new layer of commercial activity, and attract people downtown. Whatever the reasons for their implementation, skywalks have proven to be immensely popular with the public, who use them heavily regardless of weather conditions.

Despite the growing popularity of skywalks in North American cities, the literature devoted to this subject over the years has been surprisingly sparse (see literature review in chapter 1). The nature and relatively small amount of skyway-related research, together with the rapid and sometimes haphazard rate in which these systems are being constructed, leads to a basic assumption that underscored the skywalk research for this book. The assumption is that cities are adopting the skywalk as a panacea for solving downtown mobility and development problems without first assessing their positive and negative attributes and, more importantly, evaluating their long-term impact on the downtown environment. The type of research most urgently needed, therefore, should be comprehensive, examining a wide range of economic, social, planning, and design issues, as well as being comparative, encompassing systems in different cities.

The intent of this chapter is take a major step towards filling this need by examining skywalk systems in five American midwestern cities. Based on research conducted during 1985-1986, it describes and evaluates skywalks on the basis of usage, access and orientation, safety, design and aesthetics (presented in Chapter 7), economic activities, and use as a downtown redevelopment tool. The chapter is centered on five case studies: Cincinnati, Des Moines, Duluth, Minneapolis, and St. Paul (see Table 6.1). The skywalk systems in each of these cities are fairly large (minimum of 14 contiguous blocks connected), consist exclusively of skywalk linkages (none combine skywalks with below-grade linkages as some systems do), and contain numerous retail and service establishments along their corridors.

Skywalk usage

One important usage issue concerns the extent to which people use skywalks based on varying weather conditions. Pedestrian counts conducted in the mid-1970s in Minneapolis and St. Paul found that the percentage of downtown pedestrians using the skywalk ranged from an average of 46 percent in June to 76 percent in January (Podolske and Heglund 1976). The same tendency was revealed in the current study. People in the skywalk were asked whether they would be more likely to use the skywalk or the sidewalk below based on the temperature outside. For a cold day of 20 degrees Fahrenheit, at lest 96 percent in each city said they would use the skywalk.[2] A high majority still prefer the skywalk on days of 50 degrees; however, some variation exists among the cities. Even on a warm, sunny day of 80 degrees, between 71-87 percent of skywalk users in four of the cities said they would choose skywalks over the sidewalk (see Figure 6.2). The fifth city, Duluth, deviated sharply, however, as only 31 percent reported a skywalk preference. This is probably due to the fact that Duluth experiences fewer warm, sunny 80 degree days than do the other cities and its citizens are not about to waste these precious days inside a skywalk if avoidable (see Table 6.2).

In cities like Des Moines, Minneapolis, and St. Paul that experience severe cold winters, one would expect people to welcome the opportunity to walk outdoors during warm weather. However, it was found that people become so accustomed to the perceived safety, comfort, and convenience offered by skywalks that they use them regardless of the weather. Also, chosen walking routes between work and one's favorite restaurant and stores become habitual just as the route one selects to drive to work each morning seldom varies; for the majority of skywalk users, habit is a stronger force than weather conditions.

A second issue concerns the type of people who use skywalks. According to perceptions of the people surveyed, the typical skywalk user is a female white-collar worker who is more likely to earn a high income (over $50,000/year) than a modest income (under $10,000).[3] As can be seen in Table 6.3, this profile was especially predominant in Des Moines, where over 90 percent of the respondents felt that workers used skywalks more than shoppers, white-collar workers more than blue-collar, and people of high incomes more than modest income. Again the responses from Duluth varied somewhat from the other cities, as those surveyed indicated that relatively high percentages of males (42 percent), blue-collar workers (33 percent), and modest income (31 percent) used the skywalk system more. The fact that Duluth's economy is heavily dependent upon the port, manufacturing, and nearby mining may account for this difference.

One problem that this typical profile suggests is a tendency for skywalks to separate people downtown on the basis of economic class. The fact that the majority of users are perceived to be white middle-class office workers, together with the fact that the majority of structures linked by the skywalk are office buildings, upscale stores, luxury hotels, and expensive condominiums, may send signals to people with low and moderate incomes and perhaps to some minorities that they are less then welcome. One cannot help but observe the beginnings of a dual-level society downtown. It has been suggested that skywalks "can segregate a downtown by putting people of different economic levels on different physical planes." (Dillon 1982, p. 1C). Furthermore, according to Jacob and Morphew (1984), "the economic dynamics of the skyway system can lead to social stratification."

Access and orientation

Issues related to accessibility and orientation, both to and within skywalk systems, can be broken down into four problem areas: hours, street-level access, signage,and elderly and physically handicapped users. Let us begin with an examination of the hours skywalks are open to pedestrians. Since skywalk systems are seen as public or semi-public space, the same as sidewalks and streets, one might expect them to be open around the clock. However, only Cincinnati and St. Paul come close to that accomplishment. The former is open to the public 24 hours except where the skywalk passes through department stores and the latter is open from 6:00 a.m. to 2:00 a.m. during weekdays throughout the system. Hours in the other cities are inconsistent and more limited.

Most problematic is Minneapolis where hours are set block by block by the private building owners. On weekdays, opening hours vary between 6:00 a.m. and 9:30 a.m. on different blocks within the system, while buildings close their skywalk doors anywhere between 5:00 p.m. and 10:00 p.m. The inconsistencies on weekends are even greater. Several problems arise as a result of these hours. People are never quite sure exactly when the skywalk system - or parts thereof - are open, and thus may be reluctant to use it. Condominiums and apartment buildings, which proudly advertise their hook-up to the skywalk, are not always accessible by skywalk, particularly in the evening and on weekends. Finally, the existing skywalk hours discourage nighttime use. According to a 1985 editorial in the Minneapolis Star and Tribune, "if downtown Minneapolis really wants to be more than a 9-to-5 center, it must open its skyways to the same nighttime throngs it hopes will frequent its shops, restaurants, bars, theaters, concert halls, and stadiums." As expected, a smaller percentage of respondents in Minneapolis were satisfied with the system's hours than in the other cities, although a surprisingly high 87 percent were satisfied; perhaps the number would have been lower had the survey been conducted during the evening or on a weekend.

Street-level accessibility constitutes a second issue. Several studies (i.e., Lowey 1984) have commented on the difficulty of locating skywalk entrances, especially those found within buildings not detectable from the outside or, in many cases, even upon entering the building. Some users may get the feeling that they are treading on private space if they have to walk far into the building's interior to locate the skywalk. Of the five systems studied, Cincinnati's is the most accessible because many of the stairways and escalators leading to the skywalks are linked directly to the sidewalk. This enhances the system's visibility and should encourage people to use it more. The survey results, however, do not support these observations. The percentage of people who have found entrances difficulty to locate is nearly identical in both Minneapolis and Cincinnati (about 16 percent). Interestingly, only 3 percent of the respondents in Duluth reported finding entrances difficult to locate, perhaps owing to the system's relatively smaller size and fewer number of entrances.

Another problem concerns the pedestrian's ease and ability to negotiate the skywalk system once they have entered it. For those not familiar with skywalks, this might be more difficult than it sounds, as a St. Paul Pioneer Press headline indicated: "Skywalker: How Do I Get Out of Here?" While all five skywalks resemble a maze at times, the most confusing segments are where the skywalk passes through department stores. Public officials have suggested that department stores intentionally design corridors in a confusing manner so pedestrians walk by as much display merchandise as possible.

Perhaps the best way to facilitate easy skywalk orientation and movement is to provide clear and consistent directional signage. While all five systems have signs, their clarity, visibility, and consistency vary greatly. St. Paul and Des Moines both provide pedestrians with many clearly marked overhead signs that are visible from considerable distances. Cincinnati possesses seven-foot high blue-gray phylons (both at entrances and on the skywalk) that are visible but not quite as conspicuous signing. Duluth's signs are relatively small and often inconspicuous. The signage in Minneapolis is small and varies from block to block in terms of location and color.

Does clear signage and avoidance of confusing corridors through department stores translate into less people becoming disoriented within skywalks? The survey results, unfortunately, do not offer any clear-cut relationship. Des Moines offers both clear signage and obvious corridor pathways, yet almost one-third of the respondents reported getting lost or disoriented, approximately the same percentage as Minneapolis, which offers neither design consideration. The relatively low figure of 11 percent in Cincinnati may be due to the fact that parts of the system wrap around the outside of buildings rather than passing through them, thereby facilitating clearer orientation to the street. Duluth's 7 percent figure again is probably due to the system's smaller size.

Accessibility needs of the elderly and physically handicapped constitute another concern. This problem is especially notable in some of the older sections of skywalk systems. For example, parts of the Minneapolis system present the handicapped person with elevation changes connected by steps, and the steps and escalators linking Cincinnati's skywalk to the street level are equally troublesome. However, the more recently constructed links in all five systems are more sensitive to the mobility needs of the elderly and the handicapped. The Duluth system is now completely handicapped-accessible, as elevators are provided to the ground level throughout. Unfortunately, the multitude of interior doors within its skywalks still are most inconvenient to the mobility-disadvantaged, not to mention parents with strollers. In Des Moines, not only have most of the interior doors been eliminated, but the majority that remain are push-button controlled.

Safety and security

The extent to which the public will use skywalks is greatly dependent on how safe and secure people feel (Heglund 1982). This sense of security is usually based more on perceived safety then on documented crime statistics. Keeping this in mind, the majority of respondents in each of the five cities said that they felt safer on the skywalk than on the sidewalk. This was especially the case in Des Moines, where 87 percent felt safer on the skywalk, compared to 56-68 percent in the other cities. Presumably there is a direct relationship between perceived security and the overwhelmingly homogeneous profile of the typical skywalk user (female, high-income, white-collar worker).

These findings do not mean to imply, however, that security-related problems, both real and perceived, do not exist. Although skywalk crime statistics generally are not recorded as a separate category by city police departments, it is nevertheless acknowledged that skywalk crime is a concern in all five systems. The most common incidents in Minneapolis are shoplifting, purse snatching, and package theft, while in Cincinnati and St. Paul, the main problem centers on groups of teenagers who reportedly harass women and senior citizens (Jacob and Morphew 1984). Personal assaults, rapes, and rape attempts, although not frequent, appear to be on the increase, especially on the outer, less frequented skywalk segments (Morphew 1984, Lublin 1984). Some people surveyed feel less secure on skywalks, particularly during off-hours and on underfrequented outer linkages, because of the sense of being closed in, where no one could hear pleas for help, and where possible escape routes are limited. One can feel a greater sense of isolation and vulnerability on a certain skywalk corridors than on the street below.

Several remedies have been employed or proposed in an attempt to increase skywalk security. Those cities with public systems have passed ordinances that are fairly specific as to what activities are permitted on skywalks. Loitering is strictly forbidden. In Des Moines, for instance, the city code prohibits "...persons to collect, assemble or group together...or stand or loiter on any portion of the skywalk system to the hindrance or obstruction of free passage..." Furthermore, no more than two people are permitted to use the Des

Moines skywalks for peaceful picketing, public speaking, or leafleting. In St. Paul, one cannot sit, kneel, lounge, lie or otherwise recline on any skywalk floors or steps, or play a radio or tape player. Since skywalks are quasi-public places, one might seriously question the extent to which civil liberties are being violated with these restrictive ordinances (Warner 1985). A proposal in which building owners and skywalk establishments are assessed for providing additional security personnel to patrol the system was considered for use in Des Moines. In an effort to keep the skywalk clean as well as safe, the proposal suggests maintenance be increased simultaneously because "the presence of maintenance personnel are often perceived as security forces since they appear in uniform and are a continual presence." Finally, design techniques can be employed in an attempt to avoid desolate and dark corridors, blind corners, and other places that provide hiding places for would-be criminals.

Economic activities

A person experiencing skywalks for the first time often is surprised at the large number of retail and service establishments lining the inner second-level corridors, access to which, for the most part, is totally dependent on the skywalk system. As systems mature and generate increasingly greater volumes of users, they attract retailers who take advantage of the heavy pedestrian traffic flows (see Figure 6.4). A multitude of establishments can create an urban mall-type shopping environment.

The variety and availability of retail goods and services were reflected in the survey when skywalk users were asked if they used skywalk-level establishments more than street-level establishments.[4] In every city except Duluth, over 69 percent indicated that they patronize skywalk-level establishments more frequently (see Table 6.4). An astonishing 89 percent of the St. Paul respondents preferred the skywalk, giving support to those who contend it is becoming a second-story city. In contrast, only 22 percent in Duluth used stores and services on the skywalk more often than on the street, perhaps due to the relatively small number of establishments situated on the skywalk beyond the Holiday Center - an indoor shopping mall forming the hub of the system.

The economic activity on each of the five systems demonstrates considerable variation concerning the number of skywalk establishments. The number of establishments in Minneapolis and St. Paul is dramatically higher than in the other three cities. Furthermore, the distribution and concentration of establishments reveals significant differences among the five cities (see Table 6.5). At least 40 percent of the skywalk blocks in Cincinnati, Minneapolis, and St. Paul include five or more establishments, while less than 16 percent of the Des Moines and Duluth blocks contain this number. Moreover, ten establishments or more can be found on at least 20 percent of the blocks in both Twin Cities, while only one block in both Cincinnati and Duluth encompasses ten or more, and Des Moines contains no block of this size. These numbers suggest that the Minneapolis and St. Paul systems (and to a lesser degree Cincinnati's) have numerous concentrations of these establishments that not only serve pedestrian traffic but generate additional traffic flow as well. Indeed, these three systems do resemble mulitblock urban shopping malls. The same cannot be said for Des Moines and Duluth at this stage in their development.

The types of establishments found on the five systems exhibit somewhat less variation, although a few notable deviations do surface. Overall, as can be seen in Table 6.6, the majority of establishments (41 percent) can be classified as retail goods. Des Moines was the only city that deviated from this pattern, as only 25 percent of its skywalk establishments dealt with retail goods. Again, this could be related to the comparative youth of the system. Food and financial/business services each accounted for slightly over 20 percent of the overall total. The former ranged between 15 percent of all skywalk establishments in St. Paul to 26 percent in Cincinnati. The virtual absence of expensive "fine" restaurants on any of the systems (with the exception of hotel restaurants) suggest that the market is geared toward lunch-hour office workers whose time is at a premium, and that skywalk traffic during the evening is minimal. More variation exists in the financial/business services category, which is dominated by banking offices. This was the highest category in Des Moines, representing over one-third of its establishments; on the other hand, less than 13 percent of Cincinnati's establishments fall under this classification. The numbers posted by these two cities coincide with earlier findings and observations suggesting that Des Moines is dominated by white-collar office workers while Cincinnati possesses a fair share of shoppers to complement its work force. The final classification, personal services, accounts for the smallest percentage of establishments overall (12.7 percent) in four of the cities.

The most pressing problem associated with these economic trends is the tendency for skywalks to exert a negative effect on retail sales an property values on the ground floor. Nowhere is this more evident than in St. Paul. While retail sales in downtown St. Paul have continued to increase, the street-level share has steadily decreased to the point where over three-fourths of downtown retailing now takes place on the skywalk level (Lowey 1984, Wascoe 1984). This is also reflected in the fact that the 1984 average annual lease rate was $10.58 per square foot on the skywalk lever, significantly higher than the $8.90 per square foot received for first floor space (St. Paul Planning 1984). This relative scarcity of street-level economic establishments exerts a deadening effect on downtown streetlife, thus creating the perception of a lifeless downtown environment. While this problem is most severe in St. Paul, indication of this "second-story city syndrome" are evident in each of the other downtowns.

Downtown redevelopment

The potential for skywalks to spur downtown redevelopment efforts arguably has become the major impetus behind the recent planning and construction of these systems. Experts feel that "the use of skywalks as an economic tool in a downturned downtown area has overridden the climate control issue" and that, where implemented, they have "....triggered the downtown comeback." (Lublin 1984) It is difficult to gauge just how important skywalks are to the redevelopment process, but planners and public officials interviewed in each of the five cities were convinced that these overhead pedestrian systems played a crucial role.

Those who advocate the use of skywalks as a downtown redevelopment tool point to three major factors. The first concerns the importance of good pedestrian access, traditionally a hallmark of downtown viability. As Podolske (1976, p. 3) observed, "the competitive success or failure of a CBD...is often determined to a great extent by the ease and comfort of accessibility among land-use activities located within the district." A second factor revolves around changes within development practices during recent years. The emphasis towards larger projects and superblock development has yielded a greater need to help tie development together. In this regard, skywalks may serve as a unifying force. Finally, because a successful system provides an additional layer of retail/business space within the core, skywalks increase the density levels of the downtown. Compactness has long been held to be a key characteristic of a healthy CBD. These factors, together with their overall popularity with the general public, have resulted in city officials and the business community utilizing skywalk systems as a means for marketing and promoting the downtown.

Despite the frequent use of skywalks as a redevelopment tool, there are reasons to be skeptical regarding their effectiveness. The negative impact on street-level activity is one major concern. Besides the impact on street-level retailing, skywalks tend to remove other activities, and therefore people, from the sidewalk. Even if the buildings and skywalk system are pulsating with people, their absence from the street creates the impression of an inactive and barren downtown environment. There is also the question of impact on areas of the downtown not linked to the skywalk. In St. Paul, for instance, vacancy rates for buildings not connected to the system were four and one-half times higher than for buildings on the system (Middlebrook 1982). One might rightfully ask if the economic prosperity of skywalk-connected establishments is at the expense of the remainder of the downtown, and if the downtown economic growth attributed to the skywalks is merely redistribution. Finally, a skeptic might wonder why, if skywalks supposedly are so crucial to the redevelopment process, successful revitalization is occurring in downtowns across the country that do not possess such systems.

One way to examine the use of skywalks in downtown redevelopment is to observe how skywalks coordinate with other redevelopment strategies in the five cities studied. Eight major planning strategies have been implemented by cities in an attempt to revitalize their downtowns: (1) pedestrian malls, (2) indoor shopping centers, (3) special activity generators such as stadiums and convention centers, (4) improved access into and within the downtown, (5) orientation towards the office/financial sector, (6) open spaces, (7) historic preservation, and (8) waterfront development (K. Robertson 1985). There appears to be little relationship to the last two strategies, primarily because skywalks do not yet penetrate historic districts or waterfront areas in any of these cities (although several individual historical buildings have been negatively affected).

Three of the planning strategies - pedestrian malls, special activity generators, and open spaces - vary in terms of their relationship to skywalks. The dilemma with pedestrian malls is whether they complement or compete with skywalk systems. Of the five cities, only Minneapolis has such a mall. Its widely acclaimed and prototypical Nicollet Mall has had its vista rudely punctuated by several skywalks and has lost numerous establishments to buildings directly linked to the skywalk system. Nicollet Mall and the skywalk operate as two separate pedestrian systems, with no direct linkages between them. The use of skywalks to connect special activity generators is mixed. Cincinnati, Des Moines, and Duluth have skywalks linking their convention centers to major downtown stores, hotels and office buildings. Duluth's ice hockey arena is also connected, and Riverfront Stadium uses two long bridges to hook-up with downtown Cincinnati, although they fall about two blocks short of linking with the rest of the system. While Minneapolis and St. Paul both contain several large activity generators, none are connected to the skywalk system. Finally, only Cincinnati coordinates a major outdoor open space with its system. An open-sided skywalk runs the entire length of the renowned Fountain Square; not only does this segment overlook the square, but it provides direct access to it as well via outdoor escalators.

The other three downtown redevelopment strategies - indoor malls, improved access, and the office sector - are all directly related to the use of skywalks in each of the cities, although to varying degrees. Skywalks and indoor malls reciprocate each other. Skywalks serve as feeders, bringing people to the malls, while the same time indoor malls constitute the major hubs in every system except

Cincinnati. In terms of increased access, skywalks in and of themselves provide this function. Moreover, each system links to peripheral parking ramps, especially in Des Moines, which provide greater accessibility to automobile commuters. Although bus stops are located near skywalk entrances, none of the cities supplies a direct linkage. Finally, the redevelopment relationship between the growing office/financial sector should be obvious, as the most predominant land-use served by skywalk systems in each city are large office buildings.

Comparing the five systems

A review of the survey, inventory, and observations reveals similarities between the five skywalk systems. In each city, the typical skywalk user was perceived as being female white-collar worker who was more likely to earn a high income than a modest one. The vast majority of respondents in every city reported feeling safer on skywalks than on the sidewalk, never witnessing a crime on the skywalk, and not finding skywalk entrances difficult to locate or getting lost on the system. With the exception of Duluth, most people surveyed in each downtown prefer to use skywalks to sidewalks regardless of the weather outside, and report patronizing establishments on the system more than on the street. Stores selling retail goods dominate skywalk establishments in four of the five cities. Ordinances in each city prohibit individual and group behavior on skywalks that is permitted on the public street and sidewalk. The types of buildings linked by the systems primarily cater to middle -and upper-income users, thereby contributing to a physical separation of people downtown based on class. And, despite the fact that respondents overwhelmingly found skywalks to be visually attractive, simple observation clearly demonstrates that each system contains a number of design problems, particularly related to architectural integration with buildings, producing some degree of negative visual impact (see Chapter 7).

Overall, the similarities among the five systems were so prevalent that the differences were most often in terms of degree rather than general tendency. The most noteworthy differences that were uncovered, however, tend to indicate that the skywalk systems are reflective of their cities.

St. Paul

St. Paul may be described as an orderly city, dominated by governmental and institutional activity. Elements of this generalized characterization are found in its skywalks. The orderliness, for example, is captured in the standardized design pattern of the bridges, discussed in Chapter 7. This utilitarian design philosophy fosters better system-wide coordination, a more "public" appearance - like an extension of the sidewalk - and somewhat succeeds in avoiding the severe building-integration problems prevalent in other cities. It also results in skywalks that, in terms of appearance, lack interest, flair, and excitement. Not only does the uniform design of St. Paul's system contribute to this apparent lack of vitality, but the fact that people have abandoned the street for skywalks has resulted in a downtown streetscape that looks inactive and barren. The percentage of people who prefer skywalks to sidewalks on an average day (91 percent) and the number who use skywalk establishments more than street-level all provide evidence that St. Paul is rapidly turning into a second-story, indoor-oriented downtown.

Minneapolis

In stark contrast, Minneapolis seems to be more of a diverse city, one apparently more influenced by the forces of privatism and individualism. It outwardly appears "western" in its orientation, more like Denver or Dallas in looks and in attitude rather than its neighbors in the Midwest. These characteristics are reflected in the overall size and spaciousness of the system, which consists of the most skybridges, identifiable hubs, blocks connected, and blocks with ten or more establishments on the skywalk route. The fact that it is the only privately-owned system results in problems regarding hours, signage, and disorientation. It also leads to arguably the most interesting and, at times, controversial design of skybridges. The uniqueness displayed by many of the designs is indicative of the city's individuality. This factor of individuality also relates to skywalk users, who reported that they choose to use skywalks less frequently then in the other cities (except Duluth).

Des Moines

To some extent Des Moines resembles St. Paul, as it is a uniform and homogeneous city dominated by government and a conservative business sector (particularly insurance). The homogeneity is evident on the skywalk, as more survey respondents identified the typical skywalk user as a high-income, white-collar worker than in any of the other cities. This profile may explain why users also feel the safest on skywalks, and why the system contains the highest percentage of financial/business service establishments. Being the youngest of the five systems studied, Des Moines had the opportunity of learning from older systems, which undoubtedly resulted in consistent and clear signage, good handicapped accessibility, and the avoidance of corridors that pass indirectly through stores. The paucity of establishments is also a function of its relative youth.

Cincinnati

Cincinnati is the most heterogeneous and least office-oriented of the five downtowns. The comparatively high number of blue-collar, modest-income workers perceived to use the skywalk is reflective of this heterogeneity, as were the large numbers of minorities observed using the system. Cincinnati's skywalk also appeared to be the least elitist, which is not only a function of demographics but also of the system's design, which provides direct connection to the sidewalk. The city experiences the mildest climate of those studied, one consequence of which is a greater orientation towards the outside in terms of system design features. For example, it is the only system which is coordinated with a major open space (Fountain Square). It also is the only system not entirely enclosed, thereby suggesting that the skywalker's overwhelming preference for using the system rather than the sidewalk below is not necessarily a statement indicating their complete abandonment of the outdoors, as seems to be the case in Des Moines and St. Paul.

Duluth

The skywalk system that deviated from the norm most often was in Duluth, a city that is by far the smallest. Duluth is also the most rugged in terms of weather, terrain, and economic orientation (i.e., shipping, mining). The relative smallness of the city and its system helps to explain why entrances were the least difficult to locate, why people reported getting lost far less often than in other systems, and why the system has not supported many establishments beyond its major hub, the Holiday Center. Likewise, the ruggedness factor underscores several findings. According to respondent perceptions, more men, blue-collar workers, and people of modest incomes use Duluth's system than do in any of the other cities. Duluth was the only city in which people choose to use sidewalks on warm days - undoubtedly owing to the relative infrequency of warm days. People tend to patronize establishments on the street level more often than on the skywalk. Only in Duluth did a relatively high percentage of respondent (41 percent) feel as safe or safer on the street level as they did on skywalks.

Unanswered questions

As skywalks continue to be improved, current issues such as crime, disorientation, accessibility for the handicapped, and hours should become of less concern through better planning, design, and system coordination. Assuming these issues will be alleviated, certain fundamental questions remain. To what extent, for instance, does the design of urban spaces such as skywalks control our environment and the manner in which we live? Even if we invite skywalks to protect us from the natural elements, do we mind if they also control where and when we walk, what types of establishments are immediately available to us, who will be walking along side us, and what types of behavior we may participate in? Related to this issue of control is the increase in the privatization of "public" space. As a space becomes more private, either perceived or real, it becomes more restricted and the individual is able to exert less control over his/her actions, environment, and behavior. This is obviously a major concern for the privately-owned skywalks of

Minneapolis, and is most troublesome as well for the four public systems due to restrictive ordinances and the skywalk's perceived exclusiveness.

Other important questions reveal problematic dilemmas as the use of skywalks proliferates. Before they becomes widely applied and promoted as a downtown redevelopment tool, their impact on areas of the downtown not connected should be studied. Is the reported economic growth experienced by the skywalk district truly growth, or is it redistribution at the expense of other districts? A much clearer understanding of the long-term effects on streetlife created by the development of second-level downtowns must also be obtained. In terms of aesthetics, economics, and human activity, the elimination of the street level for all but vehicular functions is a very real scenario. Finally, to what extent will skywalks alter our ability and desire to sense the downtown environment and decipher its differences from other environments? Using a skywalk system is not unlike being in a suburban mall. Since most people are limited to the climatized indoors, they do not hear, small, or "feel" the downtown, and only are afforded glimpses of it when crossing skybridges. Therefore, sensually and experientially, has the skywalker even been downtown?

Endnotes

1. Besides the research methodology described in Chapter 1, this chapter draws upon a survey of approximately 100 skywalk users in each of the five cities, the purpose of which was to ascertain their usage patterns and perceptions. The surveys, which contained 19 close-ended questions, were conducted between 8:30 a.m. and 6:30 p.m. on summer weekdays during 1985. Within each system, several survey locations were used. Locations were carefully selected to ensure a reasonable cross-section of skywalk users (i.e., workers, shoppers, tourists). Potential limitations to the survey include its non-random character, the exclusive focus on skywalk users, and the absence of survey results for weekends and seasons other than summer.

2. Due to financial and time restraints, measures of skywalk usage rely on predicted behavior and perceived users. Obviously, these are only surrogates for actual behavior and users.

3. This somewhat unusual categorization of income requires explanation. It was assumed that virtually every respondent would say that middle-income users most heavily use the skywalk. Therefore, in order to determine meaningful perceived class differences, and to retain consistent use of dichotomous variables in this part of the survey, these two income extremes were selected to ascertain if people with higher than average incomes were more likely to be found on skywalks than those with more modest incomes.

4. Those surveyed were instructed not to include department stores, since they usually are multi-leveled.

Table 6.1: Background Information on American Skywalk Systems - 1985

	Cincinnati	Des Moines	Duluth	Minneapolis	St. Paul
1990 City Population	364,040	193,187	85,493	368,483	272,235
Number of Skybridges	17	19	15	28	27
Number of Contiguous Blocks Connected	14	19	15[a]	27	25
Public or Private System?	public	public	public	private	public
Major Hub	open space behind Terrace Hilton	none	Holiday Center	IDS Crystal Court	Town Square

[a] lakefront area counted as one block

Table 6.2: Skywalk Preferences Based On Temperature

(percentage preferring skywalk over sidewalk)

	Cincinnati	Des Moines	Duluth	Minneapolis	St. Paul	Overall
Cold Day (20 degrees)	100	97.1	99.0	96.0	99.0	98.2
Average Day (50 degrees)	90.1	84.5	69.0	71.7	90.9	81.3
Warm Day (80 degrees)	84.2	83.5	31.0	71.7	86.9	71.5

Notes: Survey conducted 1985
Total survey = 502 (99-102 respondents per city)

Table 6.3:Perceived Profiles of Skywalk Users -1985

(Percentage Indicating Type of Person More Likely to Use Skywalk)[a]

	Cincinnati	Des Moines	Duluth	Minneapolis	St. Paul	Overall
Women (as opposed to men)	57.4	56.3	50.0	57.6	54.2	54.2
Workers (as opposed to shoppers)	69.3	94.2	90.0	69.7	78.5	78.5
White-collar Workers (as opposed to blue-collar)	72.3	94.2	66.0	77.8	73.7	76.9
High Income (as opposed to modest income[b]	73.3	92.2	68.0	69.7	67.7	74.3

[a] Total Survey = 502 (99-102 per city)

[b] High Income = $50,000 or more; Moderate Income = under $10,000

Table 6.4: Use of Skywalk Establishments - 1985

(percentage stating they more frequently use downtown)

	Cincinnati	Des Moines	Duluth	Minneapolis	St. Paul	Overall
Skywalk Establishments	74.3	77.7	22.0	69.7	88.9	66.5
Street-Level Establishments	16.8	9.7	41.0	21.2	5.1	18.7
Both Equally	6.9	9.7	27.0	4.0	5.1	10.6

Notes: Survey conducted 1985
Total Survey = 502 (99-102 respondents per city)

Table 6.5: Distribution of Skywalk Establishments - 1985

	Cincinnati	Des Moines	Duluth	Minneapolis	St. Paul	Overall
Total Establishments	62	36	49	179	162	488
Total Contiguous Blocks	14	19	15	27	25	100
Estabs. Per Block	4.4	1.9	3.3	6.3	6.5	4.9
Blocks with 5+ Estabs.	6	3	2	14	10	35
Blocks with 10+ Estabs.	1	0	1	8	5	15
Blocks with 20+ Estabs.	0	0	1	1	3	5

note: for indoor mall-type settings, only skywalk-level establishments counted

Table 6.6: Inventory of Skywalk Establishments - 1985

City	Total Establishments	Retail Goods	Business Services	Personal Services	Food/ Drink	Other
Cincinnati OH	62	26 (41.9%)	8 (12.9%)	6 (9.7%)	16 (25.8%)	6 (9.7%)
Des Moines IA	36	9 (25%)	13 (36.1%)	5 (13.9%)	7 (19.4%)	2 (5.6%)
Duluth MN	49	22 (44.9%)	8 (16.3%)	6 (12.2%)	10 (20.4%)	3 (6.1%)
Minneapolis MN	179	75 (41.9%)	42 (23.5%)	18 (10.1%)	44 (24.6%)	0
St. Paul MN	162	68 (42%)	36 (22.2%)	27 (16.7%)	25 (15.4%)	6 (3.7%)
TOTALS	488	200 (41%)	107 (21.9%)	62 (12.7%)	102 (20.9%)	17 (3.5%)

Note: Inventory conducted during 1985

Figure 6.1: Minneapolis skywalk The oldest remaining skybridge on the five systems studies was erected over Seventh Street in 1963 (1986)

Figure 6.2: Minneapolis skywalk use. The heavy pedestrian flow on this bridge over Nicollet Mall, especially in comparison to the street below, illustrates the popularity of skywalks (1986)

7 Skywalks and downtown aesthetics

Due to their popularity with the public and their potential to spur downtown redevelopment, pedestrian skywalks may be harbingers of how the city of the 21st Century will look - an indoor-oriented and climate-controlled habitat. The proliferation of indoor shopping centers - both suburban and downtown - and domed sports stadiums, together with the rising application of downtown skywalk systems, provide clear indicators of this apparent design trend for American cities in the next century.

As presented in the previous chapter, skywalks are not panaceas for the host of ills that confront the contemporary American downtown. In fact, despite their fanfare, some critics believe that skywalks create a whole set of new problems. Some of the problems are economic, such as the drop in street-level retail sales and property values. Other problems relate to security, access and orientation, and civil liberties. One other category of problems, and the focus of this chapter, is the effect of skywalks on the attractiveness of downtown. Despite the popularity of skywalks, many critics blame them for extensive visual damage. Based on my study of skywalks in Cincinnati, Des Moines, Duluth, Minneapolis, and St. Paul, four categories of design-related impacts were uncovered: 1) lack of harmonious design with adjoining buildings, 2) inadequacies of system wide bridge design, 3) negative effects on the design at street level, and 4) blocked vistas.

Many skywalk observers (i.e., Dillon 1984, Jacob

and Morphew 1984, J. Robertson 1985, K. Robertson 1988, Maitland 1992), are concerned with the aesthetic impact these overhead walkways exert upon the downtown streetscape and environment, holding them responsible for causing extensive visual damage. For example, J. Robertson (1985, p.5) comments that skybridges "....look as if the buildings were pulled apart and a piece of the corridor system was exposed, plumbing and all." However, the average person is oblivious to criticisms of skywalk aesthetics such as this. In fact, according to the survey of skywalk users presented in chapter 6, just the reverse is true: People overwhelmingly find skywalks to be attractive! Of the 502 people surveyed in the five cities, 97 percent stated that they thought skywalks added to the visual attractiveness of their downtown; there was little deviation from city to city on this question.

Harmonious design

Making a skywalk compatible with the designs of the two buildings it connects is difficult. Often the buildings were constructed at different times, with different materials, and in different architectural styles, so a skywalk that blends attractively with one building may be totally inconsistent with the building it joins across the street (see Figure 7.1). Unfortunately, the skywalk is often an awkward appendage of both buildings. This is especially unforgivable when one building is of recent vintage, because the architects could have included the skywalk in the design from the start. In some instances a successful integration of the design styles has been achieved. However, the skywalk often seems to be an afterthought. Thus, not only do skywalks usually fail to blend with the design of both buildings they connect, frequently they do not even harmonize with the new building.

An even more delicate problem arises when the buildings are old and have historic value. Obviously the stone and brick buildings of the Nineteenth and early Twentieth Centuries were not designed to receive modern skybridges of steel and glass. The results of attaching a skybridge to them are at best awkward and at worst devastating to the building's design. Two examples of unsuccessful connections are the obliteration of the facade of the only remaining Egyptian Revival structure in Minneapolis, and the penetration of the front of St. Paul's city hall.

City officials, in part due to the outrage of local preservationists, have become increasingly concerned with this design problem. Three solutions have surfaced. First, where it is feasible, tunnels can be substituted for skywalks. This solution was used successfully at the historic city hall in Minneapolis. Here, pedestrian traffic was not disrupted because the city hall is at the end of the network. However, the level change might be a problem if designers placed a tunnel in the middle of a skywalk system. A second solution is to locate skywalks in alleys so the face of the building is not obscured. Des Moines and St. Paul have tried this technique in several places with good results (see Figure 7.2). This backdoor approach can effectively avoid difficulties in attaching new buildings as well. Finally, where alleys do not exist and tunnels are ill-advised, linking the skywalk to the side of the building can constitute a good compromise because it leaves the building's primary facade undisturbed.

Systemwide bridge design

There are two schools of thought on the overall design of bridges within a skywalk system: The St. Paul model and the Minneapolis model. St. Paul's publicly owned system consists of over thirty skybridges that all were designed to look similar. The city planning officials want the bridges to be neutral, utilitarian links that do not draw attention to themselves. The skywalks are not intended to make architectural statements but to enhance and serve the buildings. According to Carol Morphew (1984, p. 26), "ideally....bridges should be simple links to more complex structures." Cincinnati also adheres to this policy. The city's design criteria specify that bridges "should be considered as part of an integrated skywalk system with similar characteristics and image rather than isolated individual elements." Besides aesthetic considerations, an advantage to uniform design is that pedestrians more easily perceive the system as public space to be used as an extension of the sidewalk.

A contrasting approach is embodied in the

Minneapolis model in which skywalks are built and maintained privately by the property owners. Although developers must follow design standards on height, width, grade changes, transparency, and some other aspects, the guidelines are very permissive about the architectural style of bridges. The result is that Minneapolis has many different bridge designs ranging from simple and purely functional to bold and creative. The bridges are architectural statements by themselves. Certainly the skywalks in Minneapolis are more interesting than their counterparts in St. Paul, but they are also more controversial. Opinions vary widely on the attractiveness of particular bridges and how they enhance or detract from the design of the buildings they connect.

Des Moines, Duluth, and Cincinnati all lean toward the St. Paul model with bridges that tend to be uniform and neutral in style. However, each system has some unusual designs that break up the monotony. In Des Moines the most distinctive is a dark-glassed skywalk that cascades into the convention center (see Figure 7.1). Duluth has three bridges that vary from the usual pattern: An orange bridge over Michigan Street, an angled and descending bridge entering the city hall, and a 900-foot caterpillarlike bridge with many supports that connects the civic center on the lakefront with the downtown; the latter claims to be the world's longest skywalk. In Cincinnati many of the older bridges are open on the sides and wrap around the outside of buildings; most of the recent bridges, however, conform to the more conventional midstreet, enclosed, unobtrusive design. Even St. Paul has deviated from its rigid uniformity in the mid-1980s by permitting the developers of Galtier Plaza, one of the city's tallest buildings, to construct a specially designed skywalk.

Effects on the street

A skywalk system changes street-level activity and design in three stages. First, if it is successful in attracting pedestrians, the system removes people from the sidewalk. Even if the buildings and skywalks pulsate with people, the lack of human traffic on the street level creates the impression of an inactive, barren city center, thus contradicting the essence of a vital downtown.

Streetscapes without humans, no matter how well designed, are vastly incomplete.

The second stage takes place when stores and businesses relocate on the skywalk level to take advantage of heavier pedestrian traffic. In St. Paul, for example, it was estimated that over three-fourths of downtown retail business occurred on the skywalk level (Lowey 1984, Wascoe 1984). The ground level of buildings is now clearly secondary in importance, as indicated by the signs directing pedestrians to the skywalk level.

Eventually, a third stage may occur in which the design of the buildings and streetscape is oblivious to sidewalk pedestrians. The boarded-up windows on Nicollet Mall in Minneapolis and the blank street-level walls presented by Town Square and other structures built during the 1980s in St. Paul provide indications of this stage. These blank walls further deaden street life. As the 21st Century unfolds, the sidewalk may be eliminated altogether.

Blocked vistas

Skywalks blocked or cut across important vistas in each of the five cities examined. From certain vantage points, for example, views of the renowned Fountain Square in Cincinnati, the Richardsonian Romanesque city hall in Minneapolis, and the famous Aerial Bridge and scenic lakefront in Duluth are interrupted by skywalks. Perhaps the worst situation to be observed is on Locust Street in Des Moines where several skybridges cross the street, thereby intruding on a spectacularly-designed view of the Iowa State Capital which is located on a bluff overlooking the downtown (see Figure 7.4). In Minneapolis, design critics are still upset that the carefully-designed Nicollet Mall vista is now interrupted by several skybridges.

Although it may sound overly simplistic, the solution is for city planners to be more sensitive about where they locate skywalks. Rather than violating views of landmarks, skywalks can be placed elsewhere on the block or in the next block, even if that disrupts the flow of the system. This is not possible for long vistas such as Locust Street in Des Moines and Nicollet Mall in Minneapolis; because of their central location, skywalks must cross these corridors somewhere. Several cities, however, are addressing the problem of blocked vistas. In Des Moines planners have agreed that in order to preserve a view of the impressive Polk County courthouse, no skywalks will cross Court Street. Elsewhere, city officials in Seattle passed an ordinance forbidding skywalks "along designated view corridors," and they even ordered the dismantling of a skywalk that blocked a view of the Space Needle. In Edmonton, skywalks have been banned in several downtown locations to protect views and the quality of the visual environment (Maitland 1992).

Despite the loss of views caused by skywalks, it must be pointed out that the elevated walkways do create new and interesting vantage points. New second-story, midstreet perspectives offer opportunities to see the city in fresh and sometimes enlightening ways. However, these vantage points have been gained at the expense of street-level vistas, which are sometimes rudely interrupted.

Concluding remarks

Critics have a strong case for the visual damage caused by skywalks. Urban designers must be more alert to this problem in the future, and perhaps stricter design controls are needed. Yet despite the aesthetic drawbacks, people like skywalks. In bad weather even the most poorly designed skywalk is popular. Most people are unaware of the criticisms, and, in fact, most people find skywalks attractive, as the survey results presented at the beginning of this chapter clearly indicate. As architectural critic Bernard Jacob (1984, p. 36) stated: "To a very great extent our senses have been blunted and blinded by the [skywalk] system's convenience and success."

One final observation of skywalk design is that the atmosphere in skywalks is similar to that inside an enclosed shopping mall. Given the economic success of malls and the fact that downtowns are competing with the suburbs for retailing, entertainment, and business dollars, perhaps it is understandable that downtown planners are creating suburban mall-like configurations by building skywalks. One consequence is that skywalk users may experience only the climatized indoors. Pedestrians may not feel, hear, or smell the downtown, and they may see only brief glimpses of it when crossing skybridges (brief because in most systems lingering and loitering are not permitted). Therefore, in an experiential sense the typical skywalk user has not been downtown. If life downtown moves indoors, talk of visual impacts may be irrelevant because few people will be on the street to care about how skywalks affect the streetscape. Skywalks may be more that harbingers of the indoor downtown of the 21st Century, they may be a major part of it.

Figure 7.1: Des Moines skywalk This distinctive looking skywalk blends with the architecture of the convention center (left) but clashes with the building on the right (1985)

Figure 7.2: St. Paul skywalk By using the alley between two historic buildings, a skywalk was added without damaging the facades of either (1986)

Figure 7.3: St. Paul streetscape The tendency for skywalks to overwhelm and dominate the streetscape is illustrated by this view down Sixth street (1986)

Figure 7.4: Des Moines blocked vista. This skywalk rudely interrupts the vista down Locust Street of the Iowa State Capitol (1985)

8 Comparing American skywalks with Calgary's Plus 15 system

When most US planners and city officials hear the term "skywalk", they immediately think of the extensive second level pedestrian systems found in downtown Cincinnati, Minneapolis or St. Paul, not to mention numerous other smaller or less publicized US systems. Many are less aware that skywalks are also an increasingly popular planning tool in Canada, where several large Canadian cities, most notably Winnipeg, Edmonton, Calgary, and Ottawa, currently possess extensive downtown systems. Furthermore, they would undoubtedly be surprised to learn that the largest downtown skywalk system in North America is not in the USA, but in Calgary, Canada; in 1986, Calgary's 38 skybridges made it the continent's largest system. Size is not the only factor, however, that differentiates Calgary's skywalks from its American counterparts. The purpose of this chapter is to discuss the major differences between skywalks in Canadian and American cities, using Calgary's renowned Plus 15 System to represent the Canadian experience. Research on Calgary was conducted during 1986. American skywalks are represented by the research findings reported in Chapters 6 and 7 for the systems in Cincinnati, Des Moines, Duluth, Minneapolis, and St. Paul.

Comparing system objectives

As discussed in Chapter 6, the basic objectives

underscoring the development of most American systems are twofold. First, they were constructed to facilitate pedestrian mobility and safety. Skywalks permit most trips to be made in less time and more comfort, especially in bad weather, and since pedestrians are separated from vehicular traffic, improved pedestrian safety is accommodated. Second, skywalks are being used as a downtown redevelopment tool because of their ability to encourage density, to create a new layer of commercial activity on the second level and to attract people downtown. In fact, the Wall Street Journal recently reported that "the use of skyways as an economic tool in a downturned downtown area has overridden the climate-control issue." (Lublin 1984, p. 1)

While pedestrian mobility and downtown development were certainly important objectives in Calgary as well, these factors constituted only part of that city's environmentally oriented skywalk concept. When Calgary's first skywalk opened in 1970, the intent was to create a total pedestrian environment on the second level as opposed to simply providing utilitarian linkages. One of the most striking manifestations of this concept is the abundance of open space - both interior and exterior - located throughout the system. Many of the buildings have second level outdoor public plazas which comprise part of the skywalk system; most are directly accessible to the street via stairways as well. The majority of these plazas are attractively landscaped and heavily utilized, especially by office workers at lunch hour. Some open spaces, however, are barren in their appearance and are largely void of people. Several of the larger buildings also contain sizable interior open spaces, which more or less resemble indoor shopping mall complexes common to most American and Canadian downtowns. Finally, Calgary's system also relates well to several street level public plazas, most notably the Esso Building which is directly connected to the outdoor/indoor skywalk above through two well designed and very visible stairways.

The objective of climate control, which is paramount to American systems - even in cities enjoying relatively mild climates such as Cincinnati, Dallas, and Charlotte - is less important in Calgary, despite its more severe weather. Protection from the elements was not central to the vision held by the system's planners.. This is most evident not only from the outdoor open spaces linked to the system, but also from the numerous inner block corridors that are not enclosed. In fact, a few of the original skybridges are not even covered. Cincinnati is the only American system examined which is not completely enclosed; parts of its skywalk, while covered, are open to the elements on the sides.

The integration of skywalks with open space is virtually non-existent in the five systems studied in the United States. Several cities have interior spaces serving as shopping malls which constitute activity hubs; Town Square in St. Paul and the IDS Crystal Court in Minneapolis provide the best examples. However, only Cincinnati can claim to possess outdoor public spaces that relate to the skywalk. An open air skybridge runs the length of the renowned Fountain Square, providing both direct access and an overlook. Furthermore, one of the core blocks of Cincinnati's system contains an outdoor second level plaza built over the alleyways between the buildings. These were the only two examples, however, discovered in any of the five American cities, whereas open spaces serve as an integral part of the Calgary network.

Comparing system development

The manner in which Calgary's skywalk system developed has resulted in a radically different configuration from what can be observed in American cities. Its skywalks evolved incrementally, with locations of skybridges and interior walkways not being predetermined. While the evolution of each of the American systems can be described as incremental to a degree, they all conform to an overall long range plan. In Calgary, skywalks were built only when new buildings were erected downtown, regardless of location in relation to the system as a whole. Since the downtown was in the midst of a construction boom during the 1970s and early 1980s, skywalks were materializing everywhere new development took place. However, there was little coordination between the new buildings; hence the development of an integrated network of skybridges was sporadic at best.

The most significant outcome of this development pattern is that Calgary today possesses a system that is not contiguous. Most American systems modestly began with the implementation of a few bridges near the core of the downtown and gradually expanded outward over the years. For the most part, skywalks were only constructed when they could be linked to the remainder of the system, thereby maintaining a continuous, non-interrupted grade separated network. In Calgary, however, the so-called network in reality consists of over half a dozen segments that are not connected, thus forcing pedestrians to leave the skywalk for the sidewalks below should they wish to utilize different parts of the system. Eventually, perhaps, these segments will be linked together to form one integrated system; but the current non-contiguous configuration confronts the skywalk user with a confusing and inconvenient system.

Comparing transport considerations

One of the most impressive aspects of Calgary's skywalks is the way in which they interrelate with the overall downtown movement system. In the five American cities, the extent to which skywalks are coordinated with any other form of transport is limited to linkages with parking ramps, both public and private, generally situated at the peripheries of the systems. Parking ramps in Calgary serve a similar function, allowing commuters to park in one of several peripheral locations and then to utilize the elevated walkways to reach their final destination.

However, Calgary's skywalks, unlike their American counterparts, are integrated with transit as well. The city's public transit consists of buses and a recently constructed light rail transit (LRT) system, both of which serve downtown via a transit corridor implemented on centrally located Seventh Avenue. LRT station location has been coordinated with the skywalk system to permit easy transferability between the transit and pedestrian modes, as can be seen in Figure 8.1. None of the American cities have rail transit, and their bus systems are not well integrated with the skywalks.

Comparing access and orientation

One of the keys to a successful and heavily utilized grade separated pedestrian system is the ease with which people can gain access to the network and orient themselves once on the network. Access from the street level is often problematic in the American systems because entry points are usually found in the center of private buildings. Not only are the skywalk entrances not detectable from the outside, but people may get the feeling that they are treading on private space when they have to walk far into the interior of an office building or hotel to locate the skywalk. The same scenario was apparent in many places throughout the Calgary system. To make matters worse, there were few directional signs outside to give the would-be user any clue that the skywalks might be entered from within the building; to varying degrees, this was provided for in the American systems.

One remedy to this problem is to provide direct access to the skywalk from the street level, without requiring pedestrians to enter buildings. The only American city to do this is Cincinnati, which has numerous escalators and stairways leading directly to the skywalk, thereby enhancing the system's visibility and use by a wide cross-section of people. Calgary does possess numerous instances wherein stairways link the skywalk to the grade level, most often in connection with second level open spaces. Unfortunately, many of these stairways are somewhat inconspicuous, thus detracting from their effectiveness in terms of increased accessibility.

The location of entrances, together with the fact that the majority of structures linked by the systems are office buildings, up-market stores, luxury hotels, and expensive condominiums, prompts some critics to charge that skywalks are elitist. The tendency in both Calgary and the US cities is for skywalks to separate people according to economic class, in part by sending signals to people with low and moderate incomes and perhaps to some minorities that they are less than welcome. Simple observation on each of the six systems - Cincinnati being perhaps the lone exception - reveals a striking difference between the typical skywalk user and the street level pedestrian. This suggests the beginnings of a dual level downtown society in which people are physically separated by class.

Related to the question of access is the ease and ability with which pedestrians negotiate the skywalk once they have entered the system. Through maps and signing, each of the American systems attempts to facilitate clear user orientation, although their degree of effectiveness varies significantly. Probably the most successful at this are Des Moines and St. Paul, each of which provides clearly marked overhead signs that are visible from a considerable distance and indicate directions to buildings, facilities and the street below; a St. Paul newspaper headline - "Skywalker: How Do I Get Out of Here?" - suggest the helpfulness of the latter. Signs on the Duluth and Minneapolis systems are less useful due to problems of clarity, consistency, and conspicuousness.

Orientation on Calgary skywalks is a major problem because of the virtual absence of directional signs and maps and because of the inconsistency of what few signs exist. The fact that the system is not contiguous adds to the dilemma, as pedestrians have a difficult time working out how to move from one segment of the system to another. It should be noted that the City of Calgary was most aware of this orientation problem in 1986 and was in the process of installing a systemwide series of directional signs.

Comparing aesthetics

Chapter 7 described two basic schools of thought regarding the overall bridge design pattern within a system of skywalks. The first is the St. Paul model, within which all the bridges basically look the same, while the second is the Minneapolis model, where the system is comprised of a multitude of bridge designs, ranging from simple and purely functional to bridges which attempt to make bold architectural statements.

Skybridges in Calgary clearly tend to follow the Minneapolis model. The design of each bridge is determined by the owners, developers and architects of the individual buildings connected. There exists no overall design scheme nor are any design controls employed by the city. The result is a wide array of skybridges exhibiting different shapes, sizes, materials and colors. Most notable are several double decker bridges and a bridge covered with yellow glass. This diversity in design certainly makes the skywalks of Calgary and Minneapolis more interesting to view than their counterparts in the other cities. However, it also creates more controversy as opinions vary widely on the aesthetic qualities of individual bridges and the extent to which they enhance or detract from the design of downtown buildings.

In addition, there are three other major design related issues associated with skywalk systems. The first concerns the problem of older buildings being awkwardly penetrated by modern looking skybridges. This has been a particularly sensitive issue in the five American cities, especially when the buildings in question are of historical value. This has yet to constitute a problem in Calgary, however, primarily because virtually all skywalk connections are attached to newly erected buildings.

A second problem arises when skywalks block or cut across sightlines of vistas or important civic landmarks. Impressive views of monumental government buildings are rudely interrupted in Minneapolis and Des Moines (see Figure 7.4), as are attractive vistas in several American cities. Again, this does not appear to represent as serious a problem in Calgary, arguably because of the city's lack of impressive natural vistas and symbolic or historical landmarks to be found downtown. Nevertheless, the tendency for the double decker skybridges to overwhelm the streetscape and totally obstruct sightlines down Calgary's avenues does constitute a possible problem. A different perspective of this issue is offered by Calgary architect Harold Hanen (1985), who states that skywalks "arrest the boring infinity for the driver at grade level and at the upper (skywalk) level creates memorable visual experiences."

The third problem related to skywalk aesthetics revolves around the negative effect exerted on street level activity and design. The typical scenario found in Calgary and the five American cities is that skywalks tend to remove downtown pedestrians from the streets and sidewalks. Subsequently, shops and services, together with other street level activities, follow suit. This process eventually reaches the stage where the street level is clearly secondary in importance, ultimately resulting in new buildings construction and streetscape design that ignore the sidewalk pedestrian.

In downtown Calgary, the majority of the streets are lined by modern office towers with cold, blank, and uninteresting facades, reducing the sidewalk to little more than a utilitarian pathway, while the skywalks above offer shopping, activities, and public open spaces. Most of the grade level shopping is confined to a pedestrian mall implemented on Eighth Avenue. The inability to integrate the interests of small property owners and street level shopkeepers has had, according to one architectural critic, "a detrimental effect on street-level shops and on Calgary's street life generally." (Bernstein and Cawker 1982, p. 91). Downtown streetscapes which are void of people and activities, regardless of pedestrian volume in the skywalk above, present an image of a downtown that lacks vitality.

Calgary's skywalks: a contradiction in terms

An enigma emerges when trying to understand the dynamics of Calgary's skywalks system, especially in comparison to the five systems in the United States. On one hand, there are indications which suggest that the level of planning and design associated with the system is far more refined than in America. Even critics of the system point out that Calgary's skywalks have been "hailed as the most comprehensive system in North America" (Berstein and Cawker 1982, p. 89). The coordination with other forms of transport, the provision of second level open spaces, and the objective of skywalks being a component of a total downtown pedestrian environment are all indicative of this apparent commitment to skywalk planning. On the other hand, the fact that the system is not contiguous, together with the problems relating to access and orientation, suggest that incrementalism may best describe the process through which they system has developed.

One possible explanation for this enigma is related to whether Calgary's skywalks are to be considered public or private. In the five American cities, the answer to this question is more straightforward. The Minneapolis system is planned, financed, maintained, and controlled by the private building owners, while the other four systems are by and large public. In Calgary the line between public and private is not as clear. Developers of buildings to be linked to the system sign an agreement with the city requiring the developer to provide bridge supports and stairways, to ensure 24-hour public easement and to operate and maintain the skybridges, corridors, and stairways. In return, developers are given a floor density bonus.

Yet the City is able to exert little or no control over the location and design of skywalks, except between public buildings. Nor does it currently enforce the 24-hour provision even though it is clearly specified in the building titles, thus resulting in inconsistent hours and skywalk users who are unsure as to when and where the system is open. This suggests that the system may be public in name yet privately controlled in reality. As Bernstein and Cawker (1982, p. 95) conclude, the Calgary skywalks "have too often fostered an unwarranted degree of control of collective space by private interests."

One must conclude, therefore, that Calgary's skywalks are a contradiction in terms: planned incrementalism. Statements made by individuals closely associated with the system serve to support this contention. Donald Sinclair, skywalk coordinator for the City of Calgary, maintains that "the plan is a concept. Locations of bridges and walkways are not predetermined, but are worked out on a site by site basis." (1985, p. 3) Harold Hanen, a private architect and one of the original designers of the system, writes that "a plan was devised to accommodate...a system which could incrementally be extended over time as needs necessitated and means permitted." (1985, p. 2) Yet, while the lofty objective of interrelating skywalks with the overall downtown environment is exemplary, the incremental and segmented manner in which the system has evolved, together with the absence of effective controls, has served to negate this approach.

Nonetheless, American planners and city officials are encouraged to venture north to examine this distinctive system. Despite the fact that they may find the segmented characteristic of Calgary's skywalks disconcerting, there are several features which the cities contemplating expanding or initiating a system would do well to emulate.

Figure 8.1: Calgary transit/skywalk link. This skywalk links directly to the light rail transit station on Seventh Avenue (1986)

9 Skywalks v. pedestrian malls as tools for downtown development

To summarize, high levels of pedestrian activity have characterized city centers for centuries. During the past several decades, however, the volume of pedestrians on downtown streets has steadily decreased to the point where they are endangered species in many American cities. A myriad of interrelated factors have contributed to this decline. To accommodate motor vehicles, sidewalks have been narrowed to permit additional lanes of traffic. Downtown streets have become less attractive to would-be pedestrians because of insensitive building design (i.e., blank walls, parking ramps) and fewer street-level shops and activities. Downtowns have also become less walkable because distances between buildings and activity centers have increased. Finally, concerns regarding personal safety and security, in part due to the sense of isolation one often feels on downtown streets, prevent many people from walking downtown.

Not surprisingly, many of the downtown redevelopment plans prepared in recent decades have attempted to once again make the pedestrian a vital cog in the downtown. Planners realized that a safe, comfortable, and interesting pedestrian environment - the type of environment people were accustomed to in suburban shopping malls - was necessary to attract people back to the city center. Planners also needed to move people around efficiently in a comparatively dense core area, and pedestrianization schemes seemed logical. An objective of many downtown plans was to separate pedestrians from vehicular traffic, a planning approach already in wide use in Europe by the 1960s (see Chapter 2). While a wide range of strategies aimed at improving downtown pedestrianization have been implemented, the two most widely-applied large-scale pedestrian separation schemes have been skywalks and pedestrian malls. Skywalks have been implemented in dozens of downtowns throughout North America, while hundreds of cities have employed downtown pedestrian malls on both sides of the Atlantic Ocean.

This book has attempted to describe and evaluate the characteristics, problems, and potential of skywalks and pedestrian malls in American downtowns, using Swedish pedestrian streets and Calgary's Plus 15 skywalk system for comparative purposes. The purpose of this concluding chapter is to compare these two downtown pedestrian separation strategies, and to discuss the major potential benefits and problems associated with each. Obviously, the success or failure of either skywalks or pedestrian malls is partly dependent upon 1) the characteristics of the downtown contemplating their use and 2) the downtown planning objectives sought. It is not the intent of this chapter to say, in general terms, which strategy is "best", but rather to point out the relevant pros and cons to downtown planners, who, with an understanding of their city, can recommend the most appropriate alternative.

Impacts on downtown

This section will discuss the positive and negative impacts that skywalks and pedestrian malls have exerted on downtowns, based primarily on the case study cities presented in the previous chapters. The impacts to be addressed are 1) urban design, including open space considerations, 2) economic effects, particularly retail-related, 3) transportation and access, and 4) downtown image.

Urban design

The creation of attractively-designed and well-used public spaces is a common goal of most downtown plans. In the context of this goal, the differences between the two pedestrian planning strategies are significant. Pedestrian malls, which often resemble linear park-like corridors, increase the amount of public outdoor space downtown, while skywalks usually result in indoor-oriented space that is more private than public. Even if skywalks are publically owned and maintained, as they are in St. Paul, Des Moines, and Duluth, they are perceived as private because they are designed to be an integral part of the privately-owned buildings. Moreover, restrictive ordinances prevent many behaviors allowed on the street; people attempting to loiter, hand out leaflets, or make a public speech will be quickly escorted from the system by either public or private security forces. As public space becomes more privatized, either real or perceived, individuals are less able to control their behavior and environment.

To evaluate the public space qualities of skywalks and malls, it is useful to take into consideration Gehl's (1987) three levels of human activity in public spaces: 1) necessary activities - compulsory acts such as walking to work, shopping, or waiting for a bus, 2) optional activities that one usually chooses to do only if the conditions and place are inviting (i.e., strolling, sitting, sunbathing), and 3) social activities which depend on the presence of other people, such as talking, people-watching, and community/civic events. According to Gehl, the best designed public spaces are those that successfully encourage the most optional and social activities. From this perspective, skywalks cannot compare to malls because the former are designed primarily to accommodate pedestrian flow and quick-stop shopping (necessary activities); there are few opportunities to sit and people-watch on skywalk systems. Pedestrian malls certainly have greater potential to facilitate optional and social activities, but only if there are sufficient activity levels to attract people to come and stay; many downtown malls in the United States lack adequate levels of pedestrian volume needed to support these types of activities.

There are several other design-related issues of which planners should be aware. Some malls have been criticized as being overdesigned and cluttered; the abundance of landscaping and other design amenities can block sightlines of shops and provide havens for "undesirables". Critics of skywalks claim that many bridges do not integrate harmoniously with the design of buildings, especially when the buildings are older and of historic value. Skywalks blocked or cut across important vistas in each of the cities studied, although new elevated vantage points are created for the skywalker. One final design issue associated with both strategies is the proliferation of "blank walls". In skywalk cities, blank walls - such as uninteresting office building facades - work to discourage the already declining levels of street-level activity. Blank walls on pedestrian malls (i.e., windowless buildings) create dead spaces which serve to interrupt the intended continuous wall of street-oriented shops and services.

Economic impacts

Retail development is a key downtown planning objective underlying both pedestrian strategies. However, the resulting retail structure and development potential do differ. Ideally, pedestrian malls were designed to house a continuous promenade of street-oriented shops, the majority of which were small and locally-owned. A declining downtown economy, changing shopping preferences, and insensitive building design have combined to produce numerous gaps in this streetwall of retail activity. Still, the selling of retail goods in relatively small shops dominates pedestrian mall economic activity, with 61% of mall establishments engaged in this activity. The percentage of establishments involved with personal services, business services, and food/drink was considerably less (see Table 3.2); in general, these establishments tended to locate on side streets or parallel streets, where rents were usually lower.

Economic activity on skywalk systems focuses more on quick-stop convenience shopping aimed at downtown office workers. Consequently, when compared to malls, fewer skywalk establishments were retail-oriented and, as Table 6.6 demonstrates, a far greater percentage were involved in personal and business services (i.e., travel agencies, hair salons, investment services, banks). Office workers and other downtown shoppers seem to be attracted to the convenience offered, as a survey found that two-thirds of the people interviewed preferred shopping at skywalk establishments compared to street-level establishments (see Table 6.4).

The high levels of popularity and use attributed to skywalks makes it easier to attract development downtown. Mixed-use developments, with retail on the first two floors and an office building and/or hotel above, are especially attractive on skywalk systems. The resulting professionally-managed and climate-controlled environment is one in which many chain stores are very comfortable operating. Conversely, most chain stores find the retail orientation on pedestrian malls to be somewhat old-fashioned and to fall outside the range of their experience. The fact that many consumers, accustomed to going to indoor suburban malls, perceive pedestrian malls as inconvenient, uncomfortable, and less secure further erodes their potential to attract new stores.

Despite this greater commercial development potential, one negative economic impact associated with skywalks is noteworthy: The tendency for skywalks to contribute to declining street-level retailing and property values. Many street-level merchants are forced to close or relocate due to substantially reduced pedestrian traffic volume, and building owners find it increasingly difficult to lease first floor space. As shops and restaurants close, people are even less inclined to walk on the sidewalk, thereby hurting what businesses remain on ground-level. The result is a dead street. An example of this has been observed in St. Paul, where in the mid-1980s street-level establishments accounted for only 25% of downtown retail sales (St. Paul Planning 1984) and where annual lease rates were more than 20% lower per square foot for ground-level establishments compared to the skywalk level (Wascoe 1984). One therefore must ask if the economic growth attributed to skywalk development is truly growth, or is it redistribution at the expense of street-level activities (not to mention peripheral downtown buildings not connected to the system)?

Transportation impacts

Skywalk systems contribute positively to the downtown transportation pattern in two important ways. First, they usually do a better job of separating pedestrians from vehicular traffic than do pedestrian malls. By placing pedestrians on a higher level than automobiles, skywalks permit people to walk through much of the downtown without worrying about motor vehicles. Many pedestrian malls, particularly in larger cities like Denver or Minneapolis, require walkers to cross numerous streets with heavy traffic volumes. Second, skywalks can effectively link parking ramps situated on the periphery of downtown with the center, thereby encouraging suburban commuters to go directly from their cars to work without needing to step foot on the street. Peripheral parking ramps effectively serve as feeders to the overall system, although pedestrian traffic volumes are relatively light on these links themselves due to their distance from the core and relative isolation. Due to their central location, providing convenient parking in conjunction with pedestrian malls is more difficult.

However, skywalks do present some problems in terms of access and orientation that malls do not. Some people are reluctant to use these overhead walkways because entrances - often found far inside the interior of unwelcoming hotels or office buildings - are difficult to locate; moreover, some users may feel uncomfortable treading on this perceived "private space" to enter the system. Access is further reduced because many segments of the system are not open evenings and weekends. Also, it is surprisingly easy to become lost or disoriented while attempting to negotiate skywalks due to the absence of the traditional visual cues (i.e., streets, landmarks) found at ground-level.

Finally, pedestrian malls, especially those allowing transit vehicles, tend to be better linked to downtown transit systems than are skywalks. On transit malls, bus passengers are provided with direct access to the main shopping street that is unavailable to automobile drivers. Denver's fleet of electrically-powered shuttlebuses that run the length of 16th Street Mall does an outstanding job of integrating pedestrians with mass transit. Skywalks, with the notable exception of Calgary's integrated skywalk and light rail transit systems (see Figure 8.1), usually are not well-linked with transit.

Downtown image

Both pedestrian planning approaches contribute to the overall image presented by a downtown, albeit the resulting image can contrast considerably. Downtowns employing pedestrian malls can partially capture the ambience of a European city because of the mall's ability to soften the downtown environment and to make it more human-scaled and park-like. However, depending on one's point of view, malls can make the city center look antiquated and "old-fashioned"; this perception is more likely for traditional pedestrian streets than for shared or transit malls, perhaps explaining why an increasing number of traditional pedestrian streets have recently undergone conversion. The downtown image that results from a pedestrian mall is usually more related to its use than to its design. In other words, malls without people, no matter how attractively landscaped, serve to reinforce the image of a lifeless uninviting downtown.

Pedestrian skywalks, on the other hand, help to create an image of a modern and utilitarian downtown, especially when the structures being connected are new office buildings and large multi-block mixed-use developments. The emphasis on convenience, comfort, and climate-control creates a very "American-style" environment not dissimilar to that provided in an enclosed suburban shopping mall. Obviously, this image is consistent with the mainstream values held by many Americans, particularly those living in suburbia.

However, skywalks can potentially lead to two negative impacts on downtown image. First, due to their overwhelming popularity with the American public, skywalks tend to remove people and commerce from street-level, thereby exerting a deadening effect on downtown streetlife. Even if overhead walkways are pulsating with people, the relative scarcity of pedestrians, shops, and activities on the street creates a public perception of a dull and inactive downtown. Second, skywalks can potentially result in a downtown that is perceived as elitist. This planning strategy not only serves to separate pedestrians from motor vehicles, but it also tends to separates people on the basis of race and class. Most of the structures linked by skywalks house professional offices, upscale stores, luxury hotels, and expensive condominiums, thus signaling to many low and moderate income people and perhaps to some minorities that they are less than welcome. By putting people of different classes on different planes, whether by design or not, the result may be the beginnings of a dual-level downtown society.

Concluding remarks

It is difficult, if not impossible, to put forth a general recommendation as to which is the "best" pedestrian separation strategy for American downtowns. As stated at the beginning of this chapter, the recommended strategy is dependent upon the characteristics of the downtown and the planning objectives of the city in question. To determine which strategy best serves their downtown development needs, planners and other city officials should first answer the following questions with regards to their downtown.

Question #1: What type of image would you like your downtown to project? Planners must help determine if the city would be better served by a traditional street-oriented downtown possessing some European qualities or one that is more modern, utilitarian, and climate-controlled. One important consideration should be whether the suburban-oriented image projected by skywalks would benefit downtown more than an image that can offer an alternative to suburbia (pedestrian malls). The age, architecture, and present design features of the downtown will play key roles in determining the desired image.

Question #2: Which strategy would best fit with the existing downtown structure and density? A downtown characterized by high density, compactness, and low-level older buildings (e.g., most small to mid-sized eastern and midwestern cities) may be a stronger candidate for a pedestrian mall because it can integrate nicely into the street-level orientation of the core. Skywalks tend to work better when distances between buildings are greater and where taller buildings with second-floor development potential are prevalent, such as in newer downtowns that developed during the automobile era.

Question #3: How important is historic preservation to your downtown? Part of the heritage of most downtowns - especially those that were prominent prior to 1920 - is an active "main street" lined with the architecturally distinctive commercial structures. Well-used pedestrian malls have the potential to maintain this traditional downtown feature while skywalks tend to diminish downtown heritage by 1) removing pedestrians from the street and 2) destroying the facades of older buildings penetrated by skybridges. For downtowns attempting to utilize cultural tourism as an economic development tool, a skywalk system may be a hindrance.

Question #4: What type of activities do you hope

to attract to your downtown? Pedestrian malls lend themselves more to small independent shops and services that are oriented to the street, which in turn may help to invite other types of spontaneous outdoor activities (i.e., vendors, food carts, performers, people-watchers). Skywalks have the potential to attract larger indoor shopping developments which are more inviting to national chain stores that are accustomed to operating within enclosed managed shopping mall environments. In general, however, skywalks do not lead to other urban activities, either spontaneous or programmed, except for an occasional programmed event in shopping atriums connected to the system. The type and location of shopping activity differs between the two pedestrian strategies.

Planners may be tempted to implement both planning strategies simultaneously. However, with rare exceptions, this idea would prove to be unsuccessful because the two approaches are usually competitive, not complementary. Most small and medium-sized cities simply do not possess the volume of downtown foot traffic to support two separate pedestrian systems. Even in larger cities that have employed both (i.e., Minneapolis, Calgary), the mall has experienced significant activity decline as skywalk development increases; to date, these cities have not yet encouraged interblock connections between the two systems. As Whyte (1988, p. 199) stated, "the biggest problem posed by skyways....is that they will function too well. A downtown can support just so many stores and restaurants. Add another level of them and something has to give". If a city were to employ both strategies, the key would be to link them so they work together as one integrated pedestrian system (e.g., stairs or escalators directly connecting malls with skywalk).

Finally, planners should clearly understand that neither pedestrian strategy can act as a panacea for downtown revitalization. To be successful, traffic separation strategies such as skywalks and pedestrian malls must be coordinated with other development strategies (i.e., historic preservation, parking, transit, office development, cultural tourism) to help attract people back downtown.

Bibliography

Allen, H. 1981. "Downtowns Look For Malls That Work." Providence Sunday Journal (December 6), J1-J2.

American Public Transit Association. 1990. 1990 Transit Fact Book. Washington, D.C.

Anderson, K. 1988. "Fast Life Along the Skywalks." Time (August 1).

Antoniou, J. 1982. "Planning for Pedestrians." In Urban Transportation: Perspectives and Prospects, H. Levinson and R. Weant, editors. Westport, Connecticut: Eno Foundation for Transportation, 231-240.

Belkin, L. 1988. "Skywalks and Tunnels Bring New Life to the Great Indoors." New York Times (August 10): 1, 10.

Benepe, B. 1965. "Pedestrians in the City." Traffic Quarterly 19: 28-42.

Berdichevsky, N. 1984. "Gagade, The Danish Pedestrian Shopping Street." Landscape Journal 3: 15-23.

Bernstein, W. and R. Cawker. 1982. Contemporary Canadian Architecture. Toronto: Fitzhenry and Whiteside.

Boulter, D. 1988. "Downtown Denver's 16th Street Mall: A Tool for Revitalization." Center City Report 1, 4-5.

Brambilla, R. and G. Longo. 1977. For Pedestrians Only. New York: Whitney Library of Design.

Brambilla, R. and G. Longo. 1982. "Pedestrian Areas: Notes on Their Successes and Failures." In Urban Transportation: Perspectives and Prospects, H. Levinson and R. Weant, editors. Westport, Connecticut: Eno Foundation for Transportation, 240-244.

Breines, S. and W. Dean. 1974. The Pedestrian Revolution. New York: Vintage Books.

Brooks, J. 1986. "The Malling (Mauling) of Urban America." Paper presented at the annual conference of the Association of Collegiate Schools of Planning (Milwaukee, Wisconsin).

Contini, E. 1969. "Anatomy of the Mall." Journal of the American Institute of Architects 51: 42-50.

Dillon, D. 1982. "The Sky's the Limit." Dallas Morning News (January 31), 1C, 6C.

Dillon, D. 1985. "Dallas: A Case Study in Skyway Economics." Design Quarterly 129: 24-28.

Downtown Madison Partners, Inc. 1986. "Downtown Madison: Economic and Market Analysis." Madison, WI:Author.

Forusz, H. 1981. "The Cincinnati Skywalk." Contact: The Journal of Urban & Environmental Affairs 13: 335-352.

Francis, M. 1987. "The Making of Democratic Streets." In Public Streets for Public Use, A.V. Moudon, editor. New York: Van Nostrand Reinhold.

Fraser, J. 1970. "Sidewalks in the Sky." Journal of the American Institute of Architects 54: 55-56.

Freiden, B. and L. Sagalyn. 1989. Downtown Inc.: How America Rebuilds Cities. Cambridge, Massachusetts: MIT Press.

Fruin, J. 1973. "Pedway Systems in Urban Centers." Civil Engineering 43: 63-66.

Gehl, J. 1987. Life Between Buildings: Using Public Spaces. New York: Van Nostrand Reinhold.

Gehl, J. 1989. "A Changing Street Life in a Changing Society." Places 6: 8-17.

Gothenburg Planning Department. 1989. "Oversiktsplan for City i Goteborg ("Master Plan for Gothenburg's City Center")." Gothenburg, Sweden.

Grossman, L. 1987. "City Pedestrian Malls Fail to Fulfill Promise of Revitalizing Downtown." Wall Street Journal (June 17).

Gruen, V. 1964. The Heart of Our Cities. New York: Simon and Schuster.

Gunnarsson, S. and S. Lindstrom. 1969. "Existing and Planned Pedestrian Streets in Sweden 1969." Gothenburg, Sweden: Chalmers Technical Institute.

Hajdu, J.C. 1988. "Pedestrian Malls in West Germany." Journal of the American Planning Association 54: 325-335.

Hass-Klau, C. 1990. The Pedestrian and City Traffic. London: Belhaven Press.

Hennen, H. 1985. "A City For All Seasons: Weaving the Pedestrian Into the City Fabric." Livable Winter Cities Newsletter (April).

Hillman, J. 1989. "The Wonder of Wander: Walking and the Soul of Modern Society." Utne Reader (March/April): 93-94.

Houstoun, L. 1990. "From Street to Mall and Back Again." Planning 56: 4-10.

Jacob, B. 1984. "Skyways." Architecture Minnesota 10: 34-37.

Jacob, B. and C. Morphew. 1984. Skyway Typology: A Study of the Minneapolis Skyway. Washington, D.C.: AIA Press.

Klein, N. and W. Arensberg. 1974. "Auto-Free Zones:

Giving Cities Back to the People." Ekistics 219: 122-125.

Lasser. T. 1988. "The Pros and Cons of Downtown Skywalks." Urban Land 47: 2-6.

Linkoping Planning Department. 1989. "Gangtraffikmangder i Linkopings Stadskarna 1983 och 1989 ("Pedestrian Traffic Counts for Linkopings City Centre 1983 and 1989")." Linkoping, Sweden.

Lowey, B. 1984. "Skyways: Where Are They Leading Us?" Minneapolis/St. Paul 12: 66-70.

Lublin, J. 1984. "Cities are Embracing Skyways to Foster Downtown Revival." Wall Street Journal (July 11), 1.

Maitland, B. 1992. "Hidden Cities: The Irresistible Rise of the North American Interior City." Cities 9: 62-69.

Martin, J. 1990. "European Models and American Realities: A Perspective on Urban Design." Landscape 30: 36-43.

Middlebrook, W. 1982. "Skyways Turn Downtown Into Second Story." St. Paul Pioneer Press (November 22), 1, 8.

Monheim, R. 1990. "The Evolution and Impact of Pedestrian Areas in the Federal Republic of Germany." In The Greening of Urban Transport, R. Tolley, editor. London: Belhaven Press, 244-254.

Morphew, C. 1984. "The Ins and Outs of Skyways." Planning 50: 23-26.

Nordquist, S. 1984. "Gagatn-Epoken ("The Epoch of Pedestrian Malls")." Stadsbyggnad, 37-43.

Pepper, J. 1990. "Post-Earthquake Placemaking in Downtown Santa Cruz." Places 6: 92-93.

Pharaoh, T. and J. Russell. 1991. "Traffic Calming Policy and Performance: The Netherlands, Denmark, and Germany." Town Planning Review 62: 79-105.

Podolske, R. and C.T. Heglund. 1976. "Skyways in Minneapolis/St. Paul: Prototypes for the Nation."

Urban Land 35: 3-12.

Pucher, J. 1990. "A Comprehensive Analysis of Policies and Travel Behavior in the Soviet Union, Eastern and Western Europe, and North America." Transportation Quarterly 44: 441-465.

Pushkarev, B. and J. Zupan. 1975. Urban Space for Pedestrians. Cambridge, Massachusetts: MIT Press.

Roberts, J. 1990. "Summary and Conclusions for Policy." In The Greening of Urban Transport, R. Tolley, editor. London: Belhaven Press, 287-302.

Robertson, J. 1985. "Private Space in the Public Realm." Design Quarterly 129: 5-7.

Robertson, K. 1985. "Designing Downtown Redevelopment Policy: The Problem of Knowledge." Journal of Architectural and Planning Research 2: 129-139.

Robertson, K. 1988. "Pedestrian Skywalk Systems: Downtown's Great Hope or Pathways to Ruin?" Transportation Quarterly 42: 457-484.

Robertson, K. 1990. "The Status of the Pedestrian Mall in American Downtowns." Urban Affairs Quarterly 26: 250-273.

Robertson, K. 1991. "Pedestrian Streets in Sweden's City Centres." Cities 8: 301-314.

Robertson, K. 1992. "Downtown Pedestrian Malls in Sweden and the United States." Transportation Quarterly 46: 37-55.

Rotoff, B. 1981. "A Critical View of Pedestrian Malls in North America." Contact: The Journal of Urban and Environmental Affairs 13: 35-43.

Rowe, C. 1985. "I Stood in Venice on the Bridge of Sighs." Design Quarterly 129: 9-13.

Rubenstein, H. 1978. Central City Malls. New York: John Wiley.

Rudofsky, B. 1969. Streets for People. New York: Doubleday.

Sinclair, D. 1985. "The Calgary Plus 15 Pedestrian System." Paper delivered at Institute of Transportation Engineers conference (Calgary).

St. Paul Department of Planning and Economic Development. 1984. "Downtown Survey Report".

Tolley, R. 1990. "The Hard Road: The Problems of Walking and Cycling in British Cities.' In The Greening of Urban Transport, R. Tolley, editor. London: Belhaven Press, 13-33.

Untermann, R. 1990. "Why You Can't Walk There: Strategies for Improving the Pedestrian Environment in the United States." In The Greening of Urban Transport, R. Tolley, editor. London: Belhaven Press, 172-184.

Warner, S.B. 1985. "The Liberal City." Design Quarterly 129: 15-19.

Wascoe, D. 1984. "Sky Businesses Offer Life Support." Minneapolis Star & Tribune (April 2): 1M, 6M.

Whyte, W. 1988. City: Rediscovering the Center. New York: Anchor Books.

Wirtenberg, M. 1981. "Downtown Pedestrian Malls." In Shopping Centers U.S.A., G. Sternlieb and J. Hughes, editors. Piscataway, New Jersey: Center for Urban Policy Research, 229-236.

Wolfe, M.R. 1962. "Shopping Streets and the Pedestrian Rediscovered." Journal of the American Institute of Architects 37: 33-42.

Zehnpfenning, G. 1988. "Denver's 16th Street Mall." In Transit, Land Use, and Urban Form, W. Attoe, editor. Austin: Center for the Study of American Architecture, 127-134.

Zimmerman, W. 1974. "Pedestrians in the Vertical City." Real Estate Issues 4: 46-51.